CONCEPTS IN BIOCHEMISTRY

CONCEPTS IN BIOCHEMISTRY

SECOND EDITION

WILLIAM K. STEPHENSON
Earlham College

John Wiley and Sons
New York Santa Barbara
Chichester Brisbane Toronto

This book was printed and bound by Semline.
It was set in Press Roman by Graphic Technique.
The drawings were designed and executed by
John Balbalis with the assistance of the
Wiley Illustration Department.
Christine Pines supervised production.

Cover and text design by Edward A. Butler.

Library of Congress Cataloging in Publication Data:

Stephenson, William Kay.
Concepts in biochemistry.

1. Biological chemistry–Programmed instruction.
I. Title. [DNLM: 1. Biochemistry–Programmed texts.
QU18 S837c]

QD415.S785 1978 574.1'92'077 77-16205
ISBN 0-471-02002-8

Printed in the United States of America

10 9 8 7 6 5 4 3 2 1

The manuscript of the first edition of this book was prepared pursuant to a contract with the United States Office of Education, Department of Health, Education and Welfare.

PREFACE

General Purposes. This is a self-instruction textbook that covers the chemistry and biochemistry requisite for a first course in contemporary college biology. Most introductory biology courses contain considerable emphasis on biochemistry. Yet, in most cases, beginning biology students have had neither previous college chemistry nor biochemistry. Few biology texts contain adequate biochemistry coverage, and it is difficult to include this material in lectures because of the variability in student backgrounds and the time required to teach the subject. This text is designed, therefore, to bring all students to a common level of competence in biochemistry with minimal investment of student and instructional time. Particular emphasis is placed on chemical bonding and the structural formulae that are most used in biology. The levels of sophistication of the concepts covered and the pedagogy used are intended for college freshmen and sophomores.

This book is in no way a substitute for regular chemistry courses or any portion of the chemistry curriculum. Its scope is obviously not as broad, and its treatment is less thorough than conventional courses. Because it is directed toward students who may have had no previous chemical training, some concepts and treatments are oversimplified. I sincerely hope that such simplifications add clarity for the beginner and will not confuse or hinder the student who goes on to more formal chemistry training.

Assumptions as to Student Background. It is assumed that a student will bring certain skills to the task of completing this textbook: the ability to interpret data on a coordinate graph; competence in addition, subtraction, multiplication, division, and simple algebraic manipulations; and a moderate ability to read and understand English.

No previous training in chemistry is necessary to work through the material successfully. On the other hand, students who have had one or two years of secondary school chemistry or one year of college chemistry still find most of the material challenging. A cumulative self-test is included that will enable students with some experience to determine which units may be omitted and which units may be studied profitably.

Time Required. Of course, the time required to complete the text varies with the student's background and study habits. For students who have completed the book, the average total time was about 12 hours and varied between 6 and 22 hours.

Who Should Use This Text. Although this book is designed for use with the introductory college biology course, it should serve other groups equally well. A partial listing of students who should find the text helpful includes those in introductory botany and zoology courses; in advanced courses in genetics, cell physiology, microbiology, and human physiology; and in preprofessional programs such as nursing, pharmacy, medical technology, and related areas. Also those who wish to gain a background in biochemistry can do so by studying independently with this book as a guide.

Suggestions to the Instructor for Use of this Text. There are several ways in which this self-study textbook may be used. Based on my experience, the book is used most effectively when:

(1). Specific assignments are made and due dates set for each unit. The work is required and this expectation is made clear.

(2). The completed units and tests are turned in on the due dates. These are checked off by the instructor or an assistant and returned immediately. They may be spot checked for completion, but scores or grades should not be determined or recorded.

(3). Question-and-answer sessions are scheduled to coincide with the completion of the review units. These will give the students a chance to clear up any difficulties or ambiguities that may have arisen. It also will be an opportunity for the instructor to check the level of comprehension informally.

(4). Material learned in this biochemistry text is used in later components of the course. Such use will reinforce the initial learning of the material and will give this new knowledge a contextual framework. This means that the instructor must be as familiar with biochemical material as the students are.

A sample schedule for textbook use might look like the example given below. If this schedule is followed, about 5 hours of outside class time for the biochemistry work should be allowed each week.

Preceding Week	Week 1		Week 2		Week 3	
	Due Day	Units	Due Day	Units	Due Day	Units
Friday:	M	1,2, S.T.	M	9,10	M	14
Assign units,	T	3,4	T	S.T., 11	T	15, S.T.
due dates, and	W	5, S.T.	W	12	W	16
Q&A sessions.	Th	6,7	Th	13	Th	17, S.T.
	F	8, S.T.	F	S.T.	F	Cumulative Test
		Q&A session		Q&A session		Q&A session

Alternative patterns of use might include:

(1). Requiring all students to complete the text by a given date, but without compulsion. Hence there would be no due dates for specific units, no collection of completed units, and perhaps question-and-answer sessions. Outside class time is allowed by the reduction of other assignments. Students would be held responsible for the material on subsequent examinations.

(2). Assignment of the text as supplementary work, with no released time allowed. Interaction with the instructor would be at the initiation of the student. Students would be held responsible for the material on subsequent examinations.

(3). Suggesting the text as optional work. Completion of the text would not be required, and students would not be responsible for the material on examinations.

(4). Assignment of the text to be completed over the Christmas, spring, or interterm vacation periods. Opportunity for questions and discussion could be provided during the first week after the vacation.

(5). Use by individual students in independent study.

Obviously there are numerous intergrades between the patterns outlined above. The specific pattern used will ultimately depend on the judgment of each instructor with respect to the educational objectives for the course.

Following Instructions. Field tests with this book have shown that students who complete the items in writing score better on a post-test than those who do not. If students are to achieve the objectives that the book is designed to help them attain, it is important that they use the material in accordance with the instructions. The instructor can help in this by providing a course and class structure that will facilitate using the text properly. Some means of doing this were covered in the preceding section.

Some instructors will be concerned that students may peek at the correct response before recording their answers. A few students may well do this at first. They will stop, however, as soon as they realize that the purpose of the text is to teach them rather than test them, that they themselves are the only persons who will be checking for errors, and that no one cares how many mistakes they make as long as they are finally able to demonstrate that they know what is correct.

Acknowledgments. A large number of people have contributed to the development of this program. It is literally impossible to identify each one by name here. To avoid the embarrassment of omissions, let me simply express appreciation for the help of the many colleagues, students, and clerical staff who have contributed time, talent, and criticism to the manuscript during its various stages of preparation. The writing of the first edition was supported in part by a grant from the United States Office of Education to the Great Lakes College Association, under the direction of Dr. Robert

Woods Hole, Massachusetts
June, 1977

William K. Stephenson

HOW TO USE THIS BOOK

The book is divided into UNITS. Each unit covers a coherent set of material and should require less than 60 minutes for completion.

In studying each UNIT you should:

Read the CONCEPTS section, attempting to learn the vocabulary and the ideas. Do not attempt to memorize everything in this section on first reading unless you are specifically asked to do so.

Write responses to each of the QUESTIONS AND PROBLEMS. You may refer back to the CONCEPTS section, but try to respond without doing so if you can. Cover the sample response while writing your response, then check your response against the SAMPLE RESPONSE.

When you have completed the unit, you should be able to answer all questions and problems correctly without consulting the concepts section.

Discuss the material freely with other students and with instructors. Try to learn as much as you can. Be particularly careful to resolve any questions or ambiguities that are in your mind.

Reviewing should include the QUESTIONS AND PROBLEMS as well as the CONCEPTS, since important material is frequently introduced in the context of a question.

The major objectives of each unit are covered in the questions or problems whose numbers are circled; for example: Unit 1, numbers 35, 36, 38, 39, 40, and 44 are circled. These circled items are especially significant for thorough reviewing.

When you have completed the units preceding a self-test, check your capabilities with the SELF-TEST. Repeat all items missed until you reach the criterion score suggested for that test.

CONTENTS

UNIT ONE PARTICLES, ATOMS, AND MOLECULES 1

CONCEPTS

Chemical elements are the fundamental substances of the entire material universe, incuding all living matter.

Each element is comprised of exceedingly tiny units–atoms.

An atom is the smallest characteristic unit of an element.

Atoms are composed of three fundamental particles: protons, neutrons, and electrons.

Atomic Diagrams

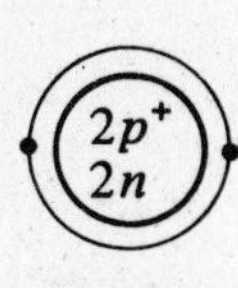

Helium

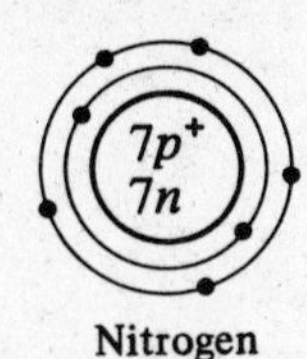

Nitrogen

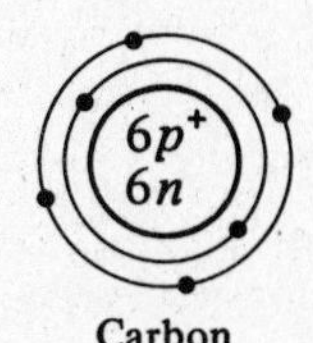

Carbon

Each fundamental particle of the atom has a distinct mass value and electrical charge:

Particle	Symbol	Electrostatic Unit of Charge	Approximate Mass	Exact Mass in Atomic Mass Units
Proton	p^+	+1	1	1.00732
Neutron	n	0	1	1.00866
Electron	e^-	-1	Negligible	0.00055

A molecule is a stable group of two or more atoms associated through bonding.

Chemical bonds are the forces that hold atoms together in a molecule.

All chemical bonding is due to electrostatic interactions of protons (p^+) and electrons (e^-).

Atomic mass = approximately the number of protons and neutrons within the nucleus.

Atomic number = number of protons within the nucleus = number of electrons around the nucleus.

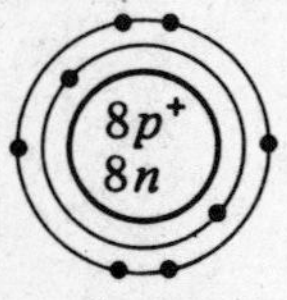

Oxygen

Atomic Name	Symbol	Atomic Number	Atomic Mass	Number of Electrons	Electrons per Shells no. 1 2 3 4	Outer shell Diagram
Oxygen	O	8	16	8	2 6	O

This material should be used for items 1-26 of Unit 1.

Table 1

Atomic Name	Atomic Symbol	Atomic Number	Atomic Name	Atomic Symbol	Atomic Number
Argon	Ar	18	Lithium	Li	3
Beryllium	Be	4	Magnesium	Mg	12
Boron	B	5	Neon	Ne	10
Calcium	Ca	20	Nitrogen	N	7
Carbon	C	6	Oxygen	O	8
Chlorine	Cl	17	Phosphorus	P	15
Fluorine	F	9	Potassium	K	19
Helium	He	2	Sodium	Na	11
Hydrogen	H	1	Sulfur	S	16

This list is incomplete and includes only those atoms which will be used subsequently. Some atomic symbols are based on the Latin names; e.g. K (kalium) and Na (natrium).

Table 2

Shell Number	Maximum Electrons per Shell
1	2
2	8
3	8 (Ar, Ca, K, Cl, S) or 10 (P)

The predominant atoms in biological molecules are C, O, H, N, P and S.

QUESTIONS & PROBLEMS

SAMPLE RESPONSES

1) The atomic number indicates the number of protons within the nucleus. How many protons are in each of the following atomic nuclei?

H____ K____ F____ C____

–H 1 K 19 F 9 C 6

2) The number of electrons characteristic of an atom is equal to the number of protons. How many *electrons* are found in each atom?

He____ Ar____ Li____ Cl____

–He 2 Ar 18 Li 3 Cl 17

3) What is the number of each of the following in a phosphorus atom?

Electrons____

Protons____

–Electrons 15

Protons 15

CONCEPTS

In the arrangement of electrons around the nucleus, each shell is filled consecutively.

Outer Shell Diagram	Atomic Number	Electrons per Shell 1	2	3	4
H	1	1			
Li	3	2	1		
P	15	2	8	5	
Ca	20	2	8	8	2
Number of electrons per filled shell:		2	8	8 or 10	8 or 10

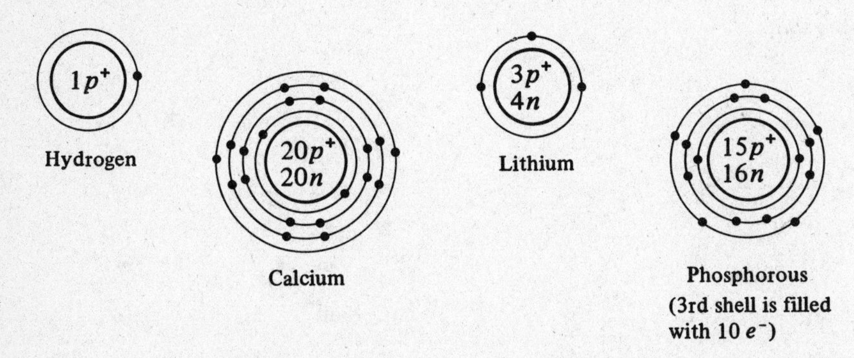

The maximum (filled) number of electrons in the outer shell yields the most stable electron configuration for an atom.

The number of electrons in the outer shells determines how atoms can combine to form molecules.

Covalent bond: outer shell electrons are shared in pairs to fill shells.

Written Formula	Structural Formula	Electron Dot Formula
H_2O	H—O—H	H:O:H
C_2H_6	H—C(H)(H)—C(H)(H)—H	H H / H C C H / H H

Double covalent bond: outer shell electrons are shared in groups of four to fill shells.

Written Formula	Structural Formula	Electron Dot Formula
C_2H_4	H₂C=CH₂ (H, H above; H, H below)	H H C::C H H
$CH_3OPO_3H_2$	H—C(H)(H)—O—P(=O)(O—H)—O—H	H / H C O P O H / H O / O / H

4) How many electrons are in each shell of a nitrogen atom?

Shell 1_____ Shell 2_____

Shell 3_____ Shell 4_____

—Shell 1 <u>2</u> Shell 2 <u>5</u>

Shell 3 <u>0</u> Shell 4 <u>0</u>

5) How many electrons are in each shell in the following atoms:

	Shell 1	Shell 2	Shell 3	Shell 4
Oxygen				
Phosphorous				
Potassium				

—Oxygen	2	6	0	0
Phosphorous	2	8	5	0
Potassium	2	8	8	1

6) When considering the combination of atoms to form molecules, we need only be concerned with the number of electrons in the outer shell.

	Number of Electrons in Each Shell (1 2 3 4)	Number of Electrons in Outer Shell	Outer Shell Diagram
H	1	1	H·
He	2	2	·H·
Li	2 1	1	Li·
Be	2 2		Be

(Electrons are first placed at each side of the symbol.)

Complete the table and write the outer shell diagram for beryllium (Be).

—2 ·Be·

7)

	Number of Electrons in Each Shell (1 2 3 4)	Number of Electrons in Outer Shell	Outer Shell Diagram
B	2 3	3	·Ḃ·
C	2 4	4	·Ċ̣·
N	2 5	5	·Ṇ̇·
O	2 6	6	·Ö̤·
F			
Ne	2 6	8	:N̈e̤:

(The third and fourth electrons are placed above and below the symbols)

—(Shells) 2 7 7 :F̤̈· (or ·F̤̈:

Complete the table and write the outer shell diagram for fluorine:

Extra electrons above 4 in the outer shell are added to produce pairs—*top and bottom first*.

8) Write the outer shell diagram for sulfur.

—·S̤̈· (2 6) (:S̈: or :S̤̈ etc. should be avoided).

9) Write the outer shell diagram for Mg.

--·Mg·

10) As atoms combine, the most stable electron configuration is one in which the outer shell contains the maximum number of electrons (Table 2). Thus, in combining, atoms may share electrons so that *the outer shell is filled*. In the diagram, circle each of the two electrons from the O atom that can be shared with the two H atoms.

– 

11) Each *pair* of shared electrons is a *covalent bond*, which "holds" the two adjacent atoms together. In the right-hand diagram circle the three covalent bonds.

Example: 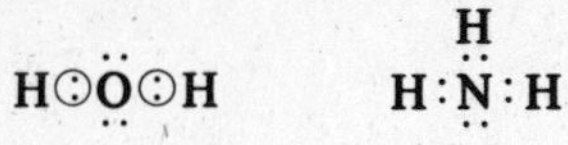

The circled electrons are shared between H and O.

– 

12) In this diagram, note that each outer electron shell is filled by the sharing of electrons.

H–2 electrons each in the outer shell
O–8 electrons in the outer shell.

For each atom in this diagram, how many electrons are in the outer shell?

Each H_____
O_____
C_____

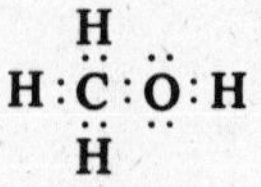

–Each H 2
O 8
C 8

13) Draw an outer shell diagram showing the combination of one C and four H atoms.

– 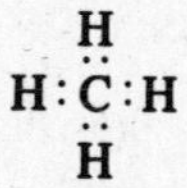

14) Circle the covalent bonds in your previous answer.

– 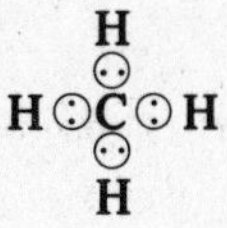

15) A stable group of two or more atoms associated through bonding is termed a molecule. Construct an outer shell diagram of a water molecule (H + H + O)

– 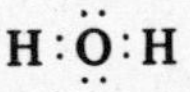

16) Two atoms of hydrogen and one atom of sulfur can combine by covalent bonding to produce a molecule of hydrogen sulfide. Diagram the outer shell structure of the molecule.

– 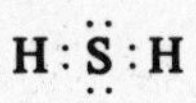

17) On the diagram of your previous answer, circle the pairs of electrons involved in covalent bonds.

– 

18) Complete the following table.

Outer Shell Diagram	Number of Possible Covalent Bonds
H·	1 H—
·Ö· (with two more dots below)	2 —O— [Each (–) represents a possible covalent bond.]
·C· (one dot above and one below)	4 —C— (with bonds above and below)
:N· (one dot below)	
·S· (two dots above, two below)	

– Outer Shell Diagram | Number of Possible Covalent Bonds

Outer Shell Diagram	Number of Possible Covalent Bonds
:N· (one dot below)	3 N— (with bonds above and below)
·S· (two dots above, two below)	2 —S—

19) Because outer shell diagrams are laborious, it is convenient to indicate each pair of shared electrons (covalent bond) by a simple dash (–) and to ignore unshared electrons.

Examples: H—S—H or H—C—O—H (with one H above and one H below the C)

This is a *structural formula*. Write the structural formula for water (H + O + H).

–H—O—H

20) Write the number of possible covalent bonds for each of the following:

H_____ O_____ S_____ N_____ C_____

Check your answer; then memorize this list for later use in writing structural formulae.

–H 1
O 2
S 2
N 3
C 4

21) How many covalent bonds can each of the following atoms form?

C_____ O_____ H_____ N_____ S_____

–C 4 O 2 H 1 N 3 S 2

22) Write the *structural formula* for methane (C + 4H) or (CH_4).

– H—C—H (with one H above and one H below the C)

23) Diagram the structural formula for methyl alcohol (CH_3OH). Begin with the first "C," then use the atoms from left to right.

– H—C—O—H (with one H above and one H below the C)

Note: CH_3 (circled) → H—C— (with one H above and one H below the C, circled); O ↓ H ↓ → O—H

24) Write the structural formula for ethyl alcohol (CH_3CH_2OH).

– H—C—C—O—H (with one H above and one H below each C)

CH_3 ↗ CH_2 ↑ O ↑ H ↑

25) In some molecules, electrons are shared in groups of *four* rather than in pairs (note circles).

H:C::C:H → double bond
H H

H:C⊙C:H
H H

On the following diagram, circle the *four* electrons that could form a double bond.

H
H:C· ·O:

—H:C::O:

26) Circle the electrons involved in double bonds.

H H
:C::C:
H H

H
:O:
H:O:P:O:H
:O:

— H:C⊙C:H (H H)

H
:O:
H:O:P:O:H
⊙
:O:

27) How many electrons are in the outer shell of each atom?

H_____
O_____
P_____

H
:O:
H:O:P:O:H or H—O—P—O—H (with H—O above P and O double-bonded below P)
:O:
Double bond

—Electrons: H 2
O 8
P 10

Note that the third shell of P is filled with 10 electrons.

28) In how many *bonds* (—) is *each* atom involved?

Each H_____
O_____
P_____

H
|
O
|
H—O—P—O—H
‖
O

—H 1
O 2
P 5

Remember: P can form 5 bonds.

Note: A double bond counts as 2 covalent bonds.

29) Complete the structural formula in the right-hand molecule.

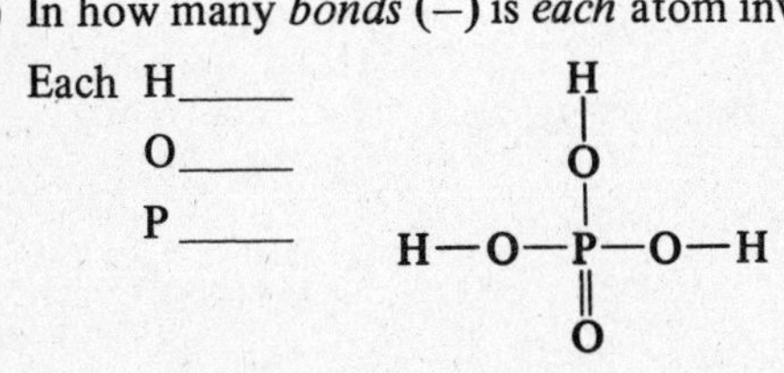

H

O

H O P O H

O

—

H
|
O
|
H—O—P—O—H
‖
O

30) Write the structural formula for formaldehyde (CH_2O). Again, begin with the C, work to the right.

— H, H \ C=O

(If you miss this, copy the correct answer).

31) Number of bonds?

Each H____ O____ C____

$$\mathrm{H_2C{=}O}$$

–H 1 O 2 C 4

32) Is this structure possible?

```
   H  H
   |  |
H—C—C—H
   |  |
   O  O—H
   |  |
   H  H
```

Explain your answer.

–No.

```
   H  H
   |  |
H—C—C—H
   |  |
   O  (O)—H
   |  |
   H  H
```

This O has 3 covalent bonds. O can form only 2 covalent bonds.

33) Write the structural formula for formic acid (HCOOH).

```
     O
    //
H—C
    \
     O—H
↑   ↑    ↑
H  COO  H
```

(If you miss this, copy the correct answer.)

34) Write the structural formula for acetic acid (CH_3COOH).

```
   H    O
   |   //
H—C—C
   |   \
   H    O—H
```

(35) What are the three fundamental particles of which atoms are composed?

–Protons, neutrons, electrons (in any order).

(36) In a calcium atom, what is the number of protons? ____ electrons?____

–Protons 20
Electrons 20

37) How many electrons are in each shell of a calcium atom?

Shell 1____ Shell 2____
Shell 3____ Shell 4____

–Shell 1 2 Shell 2 8
Shell 3 8 Shell 4 2

(38) Write the outer shell diagram for a calcium atom.

– ·Ca·

(39) Define a covalent bond.

–A *covalent bond* is formed by a *shared pair of electrons* (between atoms). (or similar response)

(40) How many covalent bonds can each of the following form?

H____ C____ O____
P____ S____ N____

–H 1 C 4 O 2
P 5 S 2 N 3

(These are the numbers of bonds the atoms *usually* form in biological systems.)

41) Diagram the structural formula for ethyl alcohol (CH_3CH_2OH).

```
   H  H
   |  |
H—C—C—O—H
   |  |
   H  H
```

42) Diagram the structural formula for formaldehyde (CH_2O).

– H–C(=O)–H

43) Diagram the structural formula for carbon dioxide (CO_2).

– O=C=O

(44) Diagram the structural formula for glycine (CH_2NH_2COOH).

– H–C(H)(N(H)H)–C(=O)–O–H (or similar response)

Take at least a 5-minute "break" before continuing on to the next unit.

UNIT TWO CHEMICAL GROUPS 2

CONCEPTS

Hydrocarbons are molecules containing only hydrogen and carbon atoms, for example, ethane and acetylene.

A *chemical group* is a submolecular group of atoms. The following chemical groups are typical constituents of many biological molecules:

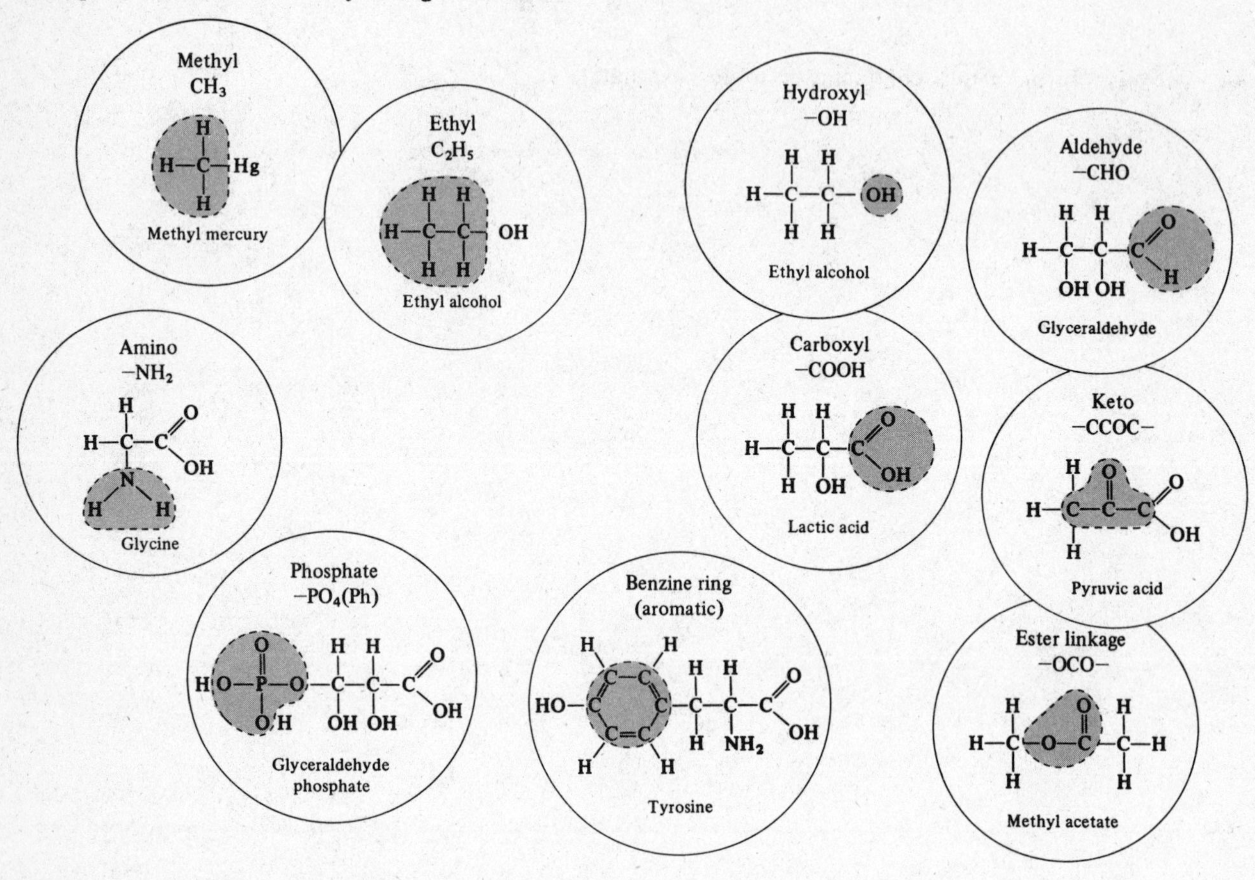

QUESTIONS AND PROBLEMS

1) Which of the following molecules are hydrocarbons?

```
      H  H                   H  H  H  H
      |  |                   |  |  |  |
A. H—C—C—H           D. H—C—C—C—C—H
      |  |                   |  |  |  |
      H  H                   H  H  H  H

      H  O     H             H        H
      |  ‖    /              |        |
B. H—C—C—N           E. H—C—C=C—C—H
      |       \              |  |  |  |
      H        H             H  H  H  H

         H
         |
         O
         |
C. H—O—P=O
         |
         O—H
```

SAMPLE RESPONSES

–A, D, and E

2) What are the distinguishing characteristics of hydrocarbons?

—Hydrocarbons contain only *hydrogen* and *carbon* atoms. (or a similar response)

3) Circle and label the methyl and ethyl groups in the following structural formulae.

```
   H          H    H        H  H        H
   |          |   /         |  |        |
 H-C-OH     H-C-N         H-C--C--O--C-H
   |          |   \         |  |        |
   H          H    H        H  H        H
```

—

```
   H              H    H        H  H        H
   |              |   /         |  |        |
(H-C-) OH      (H-C) -N       (H-C--C) -O- (C-H)
   |              |   \         |  |        |
   H              H    H        H  H        H
 Methyl         Methyl          Ethyl      Methyl
```

4) Which of the following molecules are alcohols?

```
       H                   H  H  H
       |                   |  |  |
A.  H--C--O--H      B.  H--C--C--C--H
       |                   |  |  |
       H                   H  O  H
                               \
                                H

       H  H                H  H
       |  |   //O          |  |
C.  H--C--C--C      D.  H--C--C--H
       |  |   \            |  |
       H  H    OH          O  O
                           |  |
                           H  H

       H     H             H    H
       |     |             |   /
E.  H--C--O--C--H   F.  H--C--N
       |     |             |   \
       H     H             H    H

       H  H  H  H  H
       |  |  |  |  |
G.  H--C--C--C--C--C--O--H
       |  |  |  |  |
       H  H  H  H  H
```

—A, B, D, G (C contains a carboxyl group—hence the —OH is not a hydroxyl group.)

5) What is the distinguishing characteristic of the alcohols?

—All alcohols contain one or more —O—H groups; the —OH groups must not be part of a —COOH group or phosphate group.
(—O—H = —OH = an OH group = a hydroxyl group)

6) Circle the alcohol (hydroxyl) groups in the following molecules.

```
   H            H     H  H
   |            |     |  |
 H-C-OH       H-C--O--C--C-H
   |            |     |  |
   H            H     H  H

   H  H         H              O
   |  |         |            //
 H-C--C-H     H-C--O--H    H-C
   |  |         |             \
   OH OH        H              H

   H    O
   |   //
 H-C--C
   |   \
   H    OH
```

—

```
   H              H     H  H
   |              |     |  |
 H-C-(OH)       H-C--O--C--C-H
   |              |     |  |
   H              H     H  H

   H   H           H                O
   |   |           |              //
 H-C---C-H       H-C-(O-H)     H-C
   |   |           |              \
 (OH) (OH)         H               H

   H    O
   |   //
 H-C--C
   |   \
   H    OH
```

Note that this —OH group is part of a carboxyl group; hence it is not an alcohol group.

7) Circle the aldehyde groups in the following molecules.

H—C(=O)—H   O=C(H)—CH₂—H   H—C(H)(H)—C(=O)—C(=O)—OH

—
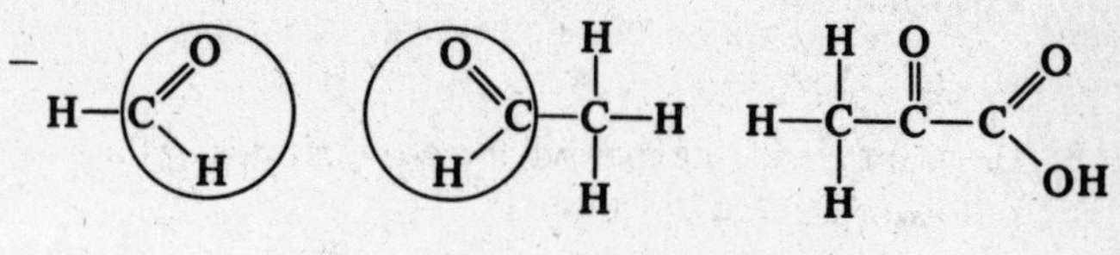

8) What are the distinguishing characteristics of an aldehyde?

—The —C(=O)—H group.

9) Which of the following are aldehydes?

A. H—C(H)(H)—C(=O)—OH

B. H—C(=O)—H

C. H—C(H)(H)—O—C(H)(H)—C(H)(H)—H

D. H—C(H)(H)—C(=O)—C(H)(H)—H

E. H—C(H)(H)—C(H)(H)—C(=O)—H

—B and E

10) Which of the following molecules are organic acids?

A. H—C(H)(H)—OH

B. H—C(=O)—OH

C. H—C(H)(H)—C(=O)—OH

D. H—C(H)(H)—C(=O)—H

E. H—C(H)(H)—O—C(H)(H)—H

F. H—C(H)(N(H)H)—C(=O)—OH

G. H—C(H)(H)—C(H)(N(H)H)—C(=O)—OH

—B, C, F, and G.

11) Write the group that is characteristic of an organic acid.

— —C(=O)—OH (This is also called a *carboxyl* group.)

12) Circle and label each group you recognize on the following molecules.

```
    H  H                    H
    |  |      O             |
H—C—C—C//            H—C—O—H
    |  |   \               |
    H  H    OH              H

    H
    |    O
H—C—C//
    |   \
    H    H
```

Answers:

```
    H  H           O            H
    |  |       C//              |
H—C—C   —    \           H—C  —  (OH)
    |  |           OH           |
    H  H                        H
  Ethyl        Carboxyl     Methyl   Hydroxyl
                (acid)               (alcohol)

    H
    |         O
H—C  —   C//
    |          \
    H           H
 Methyl    Aldehyde
```

13) Write the structural formula for each of the following groups.

Hydroxyl (alcohol)

Aldehyde

Carboxyl

Methyl

Ethyl

Answers:

```
Hydroxyl   —OH

                         O
Aldehyde            —C//
                       \
                        H

                O
Carboxyl   —C//
              \
               OH

                         H
                         |
Methyl              —C—H
                         |
                         H

            H  H
            |  |
Ethyl     —C—C—H
            |  |
            H  H
```

14) Are all molecules that contain —OH groups alcohols?

—No (The carboxyl group

```
      O
 —C//
    \
     OH
```

contains an OH group, and a phosphate group contains two OH groups, but neither are alcohols.)

15) Which of the following molecules contain a keto group?

```
      H      H                   H
      |      |                   |    O
A.  H—C—O—C—H          D.  H—C—C//
      |      |                   |    \
      H      H                   H     OH

      H                          H      H  H
      |    O                     |      |  |
B.  H—C—C//            E.  H—C—O—C—C—H
      |    \                     |      |  |
      H     H                    H      H  H

      H  O  H                    H  O  H
      |  ‖  |                    |  ‖  |
C.  H—C—C—C—H          F.  H—C—C—C—OH
      |     |                    |     |
      H     H                    OH    H
```

—C and F

16) What are the characteristics of a keto group?

```
        O                      O               O
        ‖                  —C//           —C//
–   —C—C—C—        (not      \       or      \    )
                              H                OH
```

```
                O
                ‖
Note that the C is attached to two other carbon atoms.
```

17) Which of the following contain a keto group?

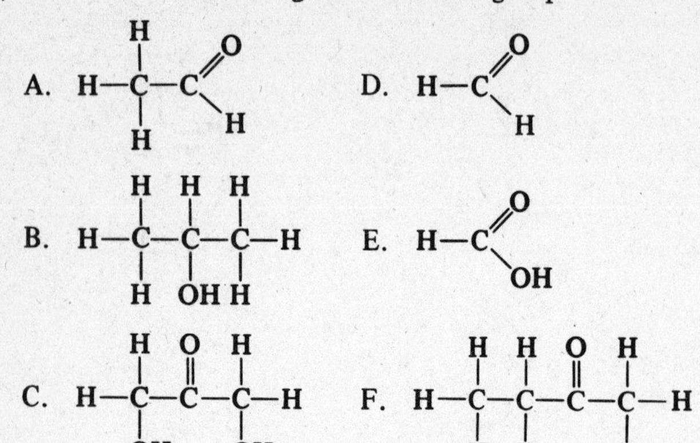

—C and F

18) Which of the following molecules contain an amino group?

```
       H     H                 H
       |    /                   \
A.  H—C—N                D.      N—O—H
       |    \                   /
       H     H                 H

                                 H
                                 |
B.  S=C=C=S              E.  H—C—C=N—OH
                                 |  |
                                 H  H

       H     O                   H  H     H
       |    //                   |  |    /
C.  H—C—C                F.  H—C—C—N
       |    \                    |  |    \
       N     OH                  H  H     H
      / \
     H   H
```

—A, C, D, and F.

19) Write the structural formula for an amino group.

```
      H
     /
 —N
     \
      H
```

20) Circle the phosphate group in the following molecule:

```
   H  O      O
   |  ||     ||
H—C—C—O—P—OH
   |         |
   H         OH
```

—

```
   H  O  ( O        )
   |  || (  ||      )
H—C—C—(O—P—OH)
   |     (  |       )
   H     ( OH       )
```

21) Write the structural formula for a phosphate group.

```
      O
      ||
 —O—P—OH
      |
      OH
```

22) Circle and label each group you recognize in the following molecules.

```
   H  O  H                 H
   |  ||  |                 |    //O
H—C—C—C—H             H—C—C
   |      |                 |    \
   H      H                 N     OH
                           / \
                          H   H

   H      O
   |      ||
H—C—O—P—OH
   |      |
   H      OH
```

—

```
  ( H )  ( O )  ( H )
  ( | )  ( || ) ( | )
(H—C—) ( C )  (—C—H)
  ( | )  ( ↓ )  ( | )
  ( H )         ( H )
Methyl   Keto    Methyl
```

```
      H
      |   ( //O )
H—(C—)(C     )
      |   ( \  )
  ( N )   ( OH )
  (/ \)
  (H   H)
 Amino    Carboxyl
          (acid)
```

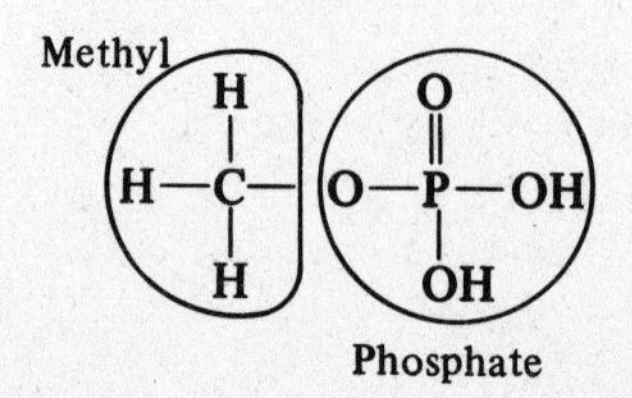

23) Write the structural formula for each of the following:

Phosphate

Keto

Amino

– Phosphate
```
          O
          ‖
    —O—P—OH
          |
          OH
```

Keto
```
      O
      ‖
   C—C—C
```

Amino
```
       H
      /
   —N
      \
       H
```

24) A carboxyl group and a hydroxyl group can combine in an *ester* linkage with the loss of H_2O.

For example,
```
     H                  O     H
     |                   \\   |
  H—C—O—H                 C—C—H
     |                   /    |
     H               H—O      H
Methyl alcohol        Acetic acid
```

Circle the H_2O atoms that could be split out as an ester linkage is formed (if you are confused, guess). Diagram the structural formula of methyl acetate (the molecule that contains the ester linkage). Circle the ester linkage.

–
```
     H                O     H
     |                 \\   |
  H—C—(OH)              C—C—H      or
     |     (H)—O       /    |
     H                      H

     H                 O     H
     |                  \\   |
  H—C—O—(H)              C—C—H
     |       H—O)       /    |
     H                       H
```

–
```
     H      O   H
     |      ‖   |
  H—C—(O—C)—C—H
     |          |
     H          H
```

25) Draw the structural formula for methyl formate. ($HCOOCH_3$)

–
```
     O       H
     ‖       |
  H—C—O—C—H
             |
             H
```

26) Which of the following are esters?

```
        H       O                  H   O   H
        |       ‖                  |   ‖   |
A.  H—C—O—C—H          C.  H—C—C—C—H
        |                          |       |
        H                          H       H

        H       H                  O       H
        |       |                  ‖       |
B.  H—C—O—C—H          D.  H—C—O—C—H
        |       |                          |
        H       H                          H
```

–A and D

27) Which of these molecules contains an aromatic group (benzine ring)?

A.

B.

–B, C, D

C.

D.

E.

28) What structure is characteristic of an aromatic compound?

–

29) The structure you just diagrammed is a benzene ring. Is this a benzene ring?

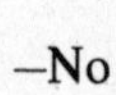

–No

30) Without consulting preceding items, write a benzene (aromatic) ring.

–

or

31) 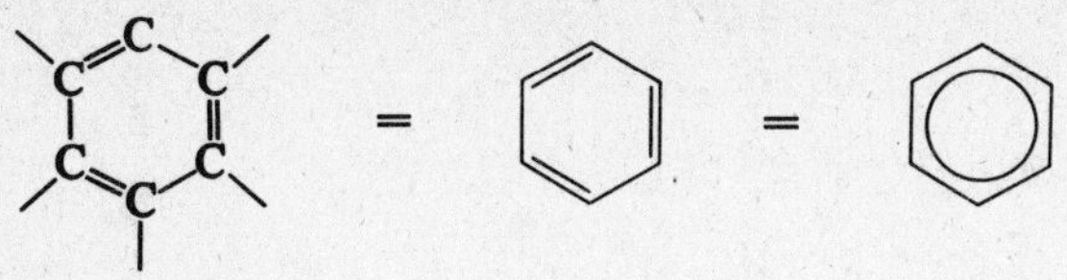 C=C ring structure = hexagon with alternating double bonds = hexagon with circle

Which of the following are aromatic compounds?

–B and E

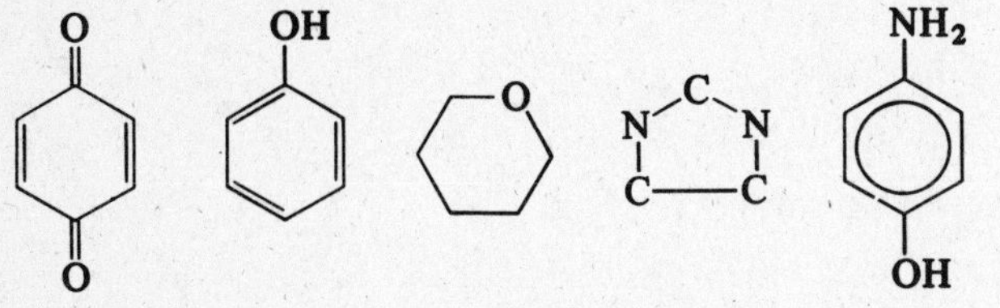

A. (ring with two =O) B. (benzene ring with OH) C. (ring with O) D. (five-membered ring with N, C, N, C, C) E. (ring with circle, NH_2 and OH)

32) Group review: Match the following:

Alcohol________
Aldehyde________
Amino________
Aromatic________
Carboxyl________
Ester linkage____
Ethyl________
Keto________
Methyl________
Phosphate________

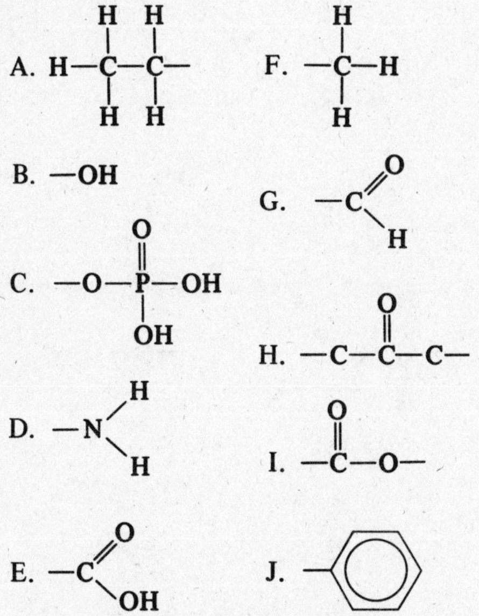

A. H—C—C— (each C with H above and below)

B. —OH

C. —O—P(=O)(OH)—OH

D. —N(H)H

E. —C(=O)OH

F. —C—H (C with H above and below)

G. —C(=O)H

H. —C—C(=O)—C—

I. —C(=O)—O—

J. —(benzene ring)

–Alcohol B
Aldehyde G
Amino D
Aromatic J
Carboxyl E
Ester linkage I
Ethyl A
Keto H
Methyl F
Phosphate C

33) Label all the groups you know in the following molecules.

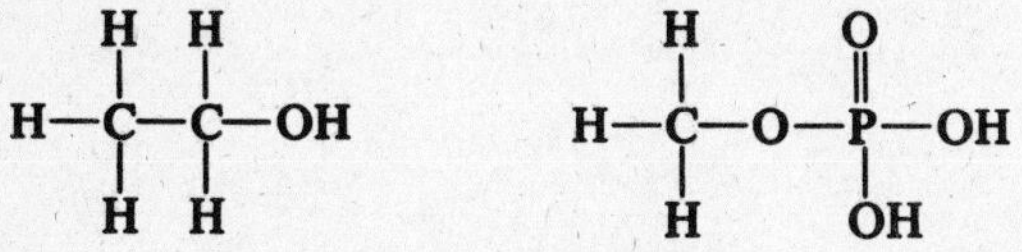

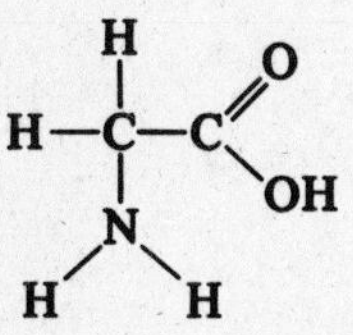

H—C—C—OH (each C with H above and below)

H—C—O—P(=O)—OH (C with H above and below; P with OH below)

H—C(H)(N(H)H)—C(=O)OH

–

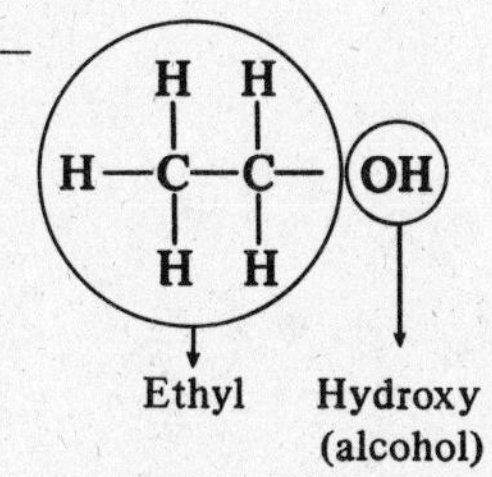

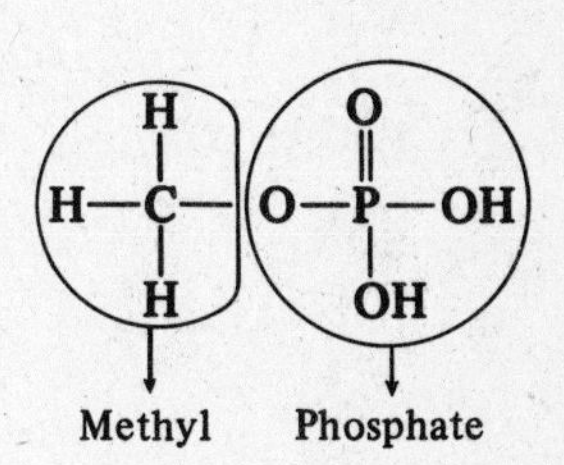

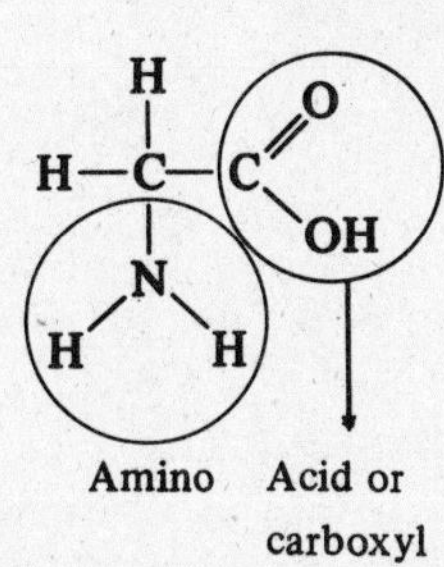

(34) Diagram each of the following groups.

Ethyl

Amino

Phosphate

—

```
        H  H
        |  |
Ethyl  —C—C—H
        |  |
        H  H
```

```
              H
             /
Amino    —N
             \
              H
```

```
                   O
                   ||
Phosphate   —O—P—OH
                   |
                   OH
```

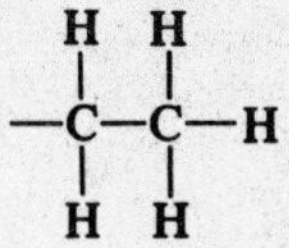

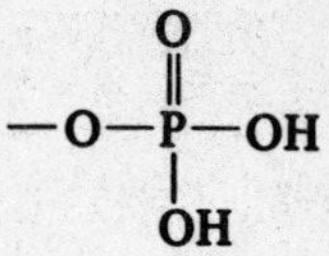

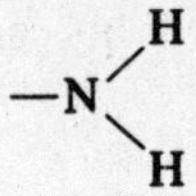

(35) Diagram each of the following groups.

Keto

Aromatic

Carboxyl (acid)

Ester linkage

—

```
           O
           ||
Keto   —C—C—C—
```

```
              \    /
               C—C
              //   \\
Aromatic  —C       C—   or  (hexagon with alternating double bonds)  or  (hexagon with inscribed circle)
              \    /
               C=C
              /    \
```

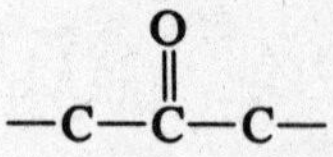

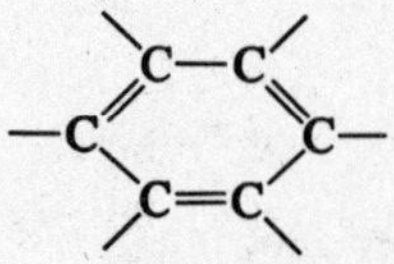

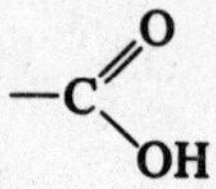

```
              O
             //
Carboxyl  —C
             \
              OH
```

```
                   O
                  //
Ester linkage  —C
                  \
                   O—
```

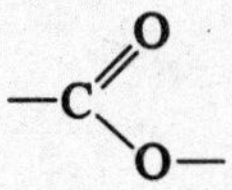

(36) Diagram each of the following groups:

Methyl Hydroxyl (alcohol) Aldehyde

—

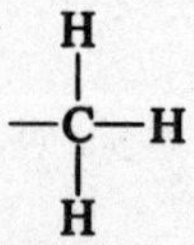

```
  H                               O
  |                              //
—C—H          —O—H            —C
  |                              \
  H                               H
Methyl         Alcohol         Aldehyde
```

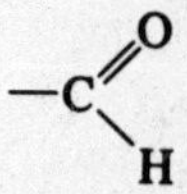

Take at least a 5-minute break before continuing on to the self-test.

SELF-TEST UNITS 1&2

(Criterion score 55 points: total of 58 points)

Retention of material you have learned is a vexing educational problem! This self-test covers the main points of the first two units and will help you to check your level of retention. Work through it just as you have worked through the preceding units, with the exception of waiting to check your responses until you have finished the entire test. Use the Self-Test Key to check your responses. Whenever you judge your response to be inadequate check back over the items indicated in the response space. Review carefully to be certain that you understand the principles involved. The self-tests should communicate to you the level of expectation for retention and should help you to know at what level subsequent material should be learned.

You should repeat all items that you judge to be incorrect until you attain the criterion score.

(5 points) 1) For calcium (atomic number = 20) fill in each of the following numbers:
Protons in nucleus
Electrons in shell No. 1
Electrons in shell No. 2
Electrons in shell No. 3
Electrons in shell No. 4

(3 points) 2) Write the *outer shell diagram* for fluorine (atomic number = 9)

(5 points) 3) How many covalent bonds can each of the following atoms form?
C____ O____ H____ N____ S____

(5 points) 4) Write the structural formula for pyruvic acid (CH_3COCO_2H).

(10 points) 5) Group review: match the following:

Alcohol________
Aldehyde________
Amino________
Aromatic________
Carboxyl________
Ester linkage________
Ethyl________
Keto________
Methyl________
Phosphate________

A. H—C(H)(H)—C(H)(H)—
B. —OH
C. —O—P(=O)(OH)—OH
D. —N(H)H
E. —C(=O)OH
F. —C(H)(H)—H
G. —C(=O)H
H. —C—C(=O)—C—
I. —C(=O)—O—
J. —C₆H₅ (benzene ring)

(9 points) 6) Diagram each of the following groups:

Ethyl

Amino

Phosphate

(12 points) 7) Diagram each of the following groups:

Keto

Aromatic

Carboxyl (acid)

Ester linkage

(9 points) 8) Diagram each of the following groups:

Methyl

Hydroxyl (alcohol)

Aldehyde

(58 points total)

SELF-TEST KEY UNITS 1&2

(Criterion score 55 points: total of 58 points)

______

(5 points) 1) 20 (1 point) No. 1–2 (1) No. 2–8 (1) No. 3–8 (1) No. 4–2 (1)
(If you miss this, review Unit 1, Items 1-8.)

(3 points) 2) $:\ddot{\underset{\cdot\cdot}{F}}\cdot$ (1 point off for each error, but not to exceed -3) *(If you miss this, review Unit 1, Items 9-12.)*

(5 points) 3) C 4 O 2 H 1 N 3 S 2 (1 point each)
(Review: Unit 1, Items 13-24)

(5 points) 4)

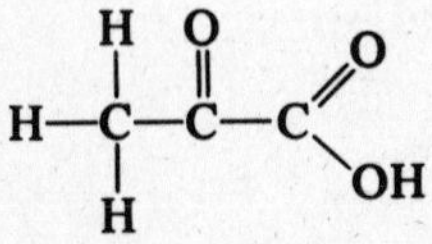

```
    H  O
    |  ‖   O
H—C—C—C⁄⁄
    |      \
    H       OH
```

(1 point off for each error, but not to exceed -5.)
(If you miss this, review Unit 1, Items 24-40)

(10 points) 5)

Alcohol	B	Ester linkage	I
Aldehyde	G	Ethyl	A
Amino	D	Keto	H
Aromatic	J	Methyl	F
Carboxyl	E	Phosphate	C

(1 point each) *(Review: Unit 2, appropriate sections)*

(9 points) 6) (1 point off for each error, but not to exceed -3 per item.)

Ethyl

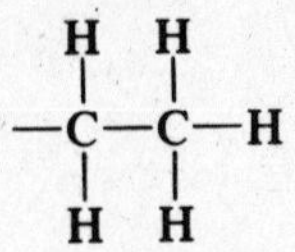

```
  H  H
  |  |
—C—C—H
  |  |
  H  H
```

(Review: Unit 2, Items 1-3)

Amino

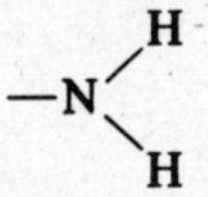

```
     H
    /
—N
    \
     H
```

(Review: Unit 2, Items 18-19)

Phosphate

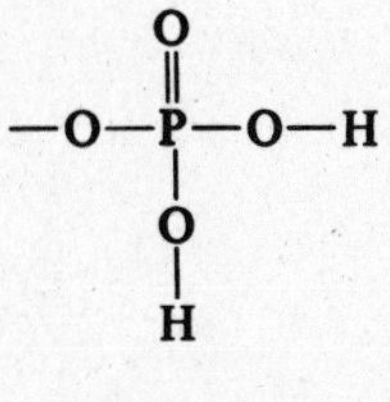

```
       O
       ‖
—O—P—O—H
       |
       O
       |
       H
```

(Review Unit 2, Items 20-21)

(12 points) 7)

Keto

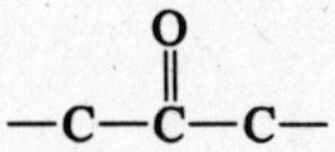

```
      O
      ‖
—C—C—C—
```

(Review: Unit 2, Items 15-17)

Aromatic

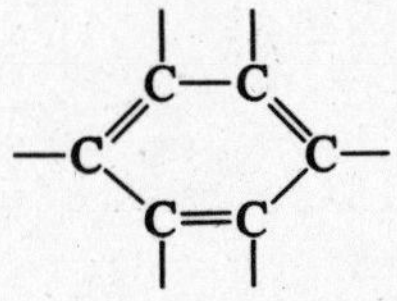

```
      |  |
      C—C
     ⁄⁄    \\
—C          C—
     \     ⁄
      C=C
      |  |
```

or

—⌬ or —⌬ (with circle)

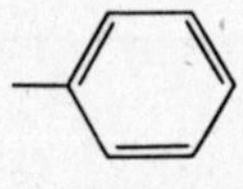

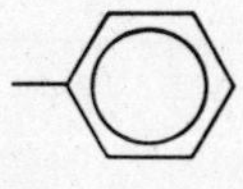

(Review: Unit 2, Items 27-31)

Carboxyl $-C{\overset{\displaystyle O}{\diagup\!\!\diagup}}\!\!\diagdown OH$ (*Review: Unit 2, Items 10 and 11*)

Ester linkage $-C{\overset{\displaystyle O}{\diagup\!\!\diagup}}\!\!\diagdown O-$ (*Review: Unit 2, Items 24-26*)

(9 points) 8)

Methyl $-\underset{H}{\overset{H}{\underset{|}{\overset{|}{C}}}}-H$ (*Review: Unit 2, Items 1-3*)

Alcohol $-O-H$ (*Review: Unit 2, Items 4-6*)

Aldehyde $-C{\overset{\displaystyle O}{\diagup\!\!\diagup}}\!\!\diagdown H$ (*Review: Unit 2, Items 7-9*)

(58 points total)

Take at least a 5-minute break before going on to the next unit.

UNIT THREE IONS, IONIC BONDS, AND CRYSTALS 3

CONCEPTS

Electrostatic *attraction* occurs between opposite electrical charges: + and -.

Electrostatic *repulsion* occurs between identical charges: - and -, or + and +.

An ion is an atom that, through electron transference, has lost or gained electrons in its outer shell (or a group of atoms in which one atom has gained or lost outer shell electrons):

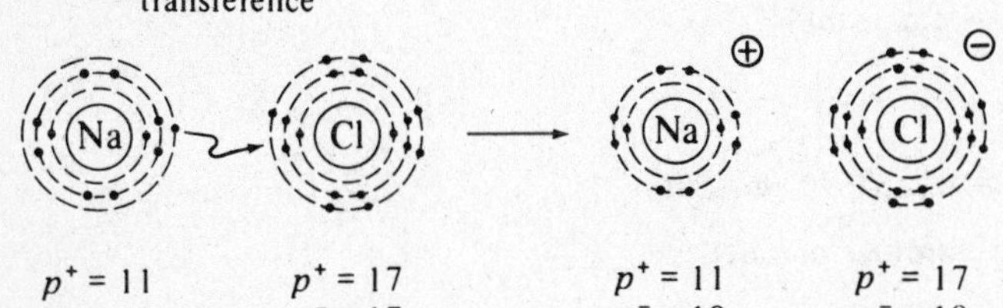

An ion usually has gained or lost electrons to form a filled outer shell.

In an ion, the total number of protons (p^+) is greater than or less than (but never equal to) the total number of electrons (e^-).

An anion is a negatively charged ion; examples: Cl^-, $SO_4^=$ ($p^+ < e^-$).

A cation is a positively charged ion; examples: Na^+, Ca^{++} ($p^+ > e^-$).

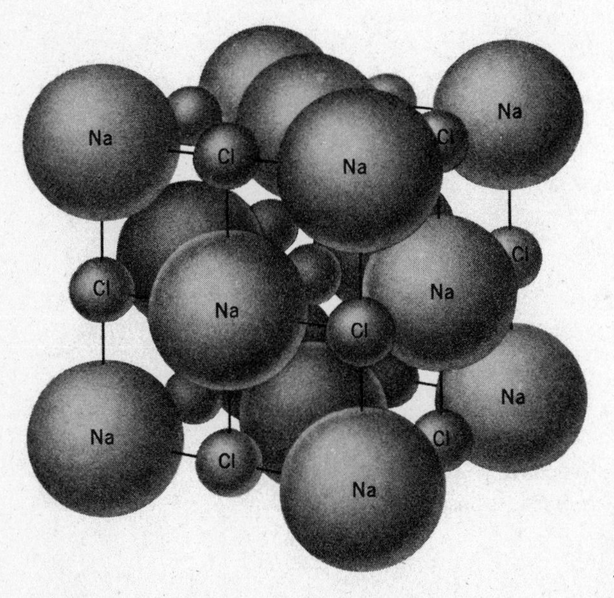

Diagram of the Association of Ions in a Crystal of Salt (Sodium Chloride). (From G. C. Stephens and B. B. North, *Biology*, John Wiley & Sons © 1974.)

Ionic bond: the aggregation of two or more ions due to electrostatic attraction. For example, Na^+ and Cl^- in a crystal of salt:

Salts are ionic compounds or ionic solids.

A salt is an ionic compound in which the cation is usually a metallic ion, for example, Na^+, K^+, Ca^+, Mg^{++}, Fe^{++}, Fe^{+++}

Dissociation: the breakdown of an ionic compound in water, resulting in the formation of free ions.

QUESTIONS AND PROBLEMS / SAMPLE RESPONSES

1) What electrostatic charge (+ or −) do you associate with electrons? _____ protons? _____ Which of the following electrostatic charge pairs are attracted to one another? (++) (+−) (−−)?

—Electrons __−__ protons __+__

(+−) (Opposite charges attract; like charges repel.)

2) A covalent bond is due to electrostatic attraction. For example, in water H:Ö:H, what part of the hydrogen atom (nucleus or electron) is attracted to the oxygen nucleus?

—Electron (The − electron is attracted to the + oxygen nucleus. Thus, the H atom is bonded to the O atom by the electrostatic attraction between its electron and the + oxygen nucleus.)

3) In a water molecule, to what molecular component is the hydrogen nucleus attracted?

—The electrons (the + hydrogen nucleus is attracted to the − electrons)

④ List the following for a sodium atom.

Number of protons_____.

Number of electrons_____.

Is the sodium atom electrostatically neutral; (the total + charge of the nucleus equals the total − charge of the electrons) (protons equals electrons)?

—Number of protons __11__

Number of electrons __11__

—Yes (+ = −)

5) In general, the outer shell of an *ion* is either empty or complete when compared to an atom.

Outer Shell Diagrams

Atoms	Ions	Ionic Charge
Na·	Na	+1
·Ca·	Ca	+2
:C̈l·	:C̈l:	−1
·Ö·	:Ö:	−2

Check the table carefully to determine the characteristics of ions. Is the following an ion? If so, what is its charge?

:F̈:

—Yes

−1

⑥ What is the general characteristic of the ions in the preceding item (when compared to the atoms)?

—The ions have gained or lost electrons and bear a charge. (or similar response)

7) The sodium ion is written Na^+. Ca ion = Ca^{++}.

Write the symbol for the Mg ion.

Write the symbol for the K ion.

—Mg^{++} and K^+

8) Outer shell diagram

How many electrons can Cl accept to form an ion (complete the outer electron shell)? Write the symbol for the Cl ion.

–One
Cl^-

⑨ Write the symbol for each of the following ions:

Na_____ Mg_____ K_____
Cl_____ Ca_____ F_____

–Na Na^+ Mg Mg^{++} K K^+
Cl Cl^- Ca Ca^{++} F F^-

⑩ Cation = ⊕ ion Pronounced: CAT´ ION
Anion = ⊖ ion

Pronounce cation and anion out loud three times each.

Label each as a cation or anion.

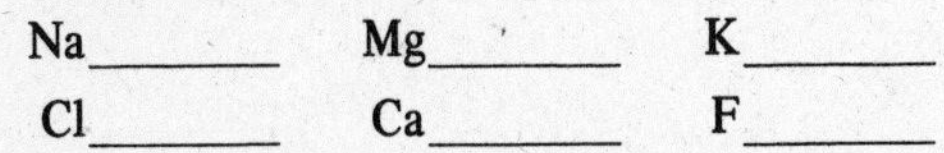

Na_______ Mg_______ K_______
Cl_______ Ca_______ F_______

–Na cation Mg cation K cation
Cl anion Ca cation F anion

11) A *salt* is a general term for an *ionic* compound. Of what units is a salt composed?

–Ions

12) Most salts have metallic (e.g., sodium, potassium, magnesium, iron, calcium) cations. Circle the metallic ion in each of these salts.

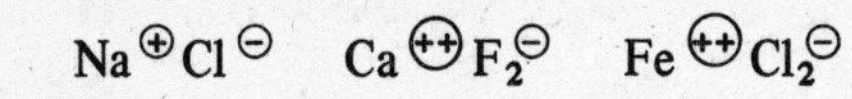

$Na^{\oplus}Cl^{\ominus}$ $Ca^{++}F_2^{\ominus}$ $Fe^{++}Cl_2^{\ominus}$

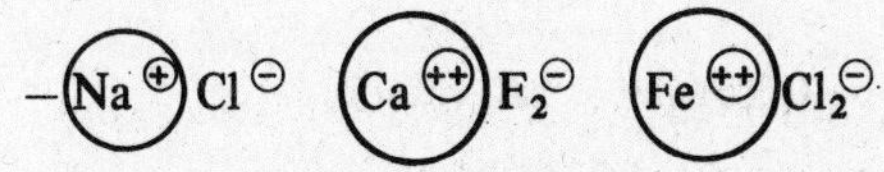

⑬ How would you define a salt?

–A salt is composed of ions; the cation is usually a metallic ion. (or similar response)

14) For convenience, an ionic bond is usually diagrammed to indicate the attraction of opposite charges. Thus, sodium formate could be written:

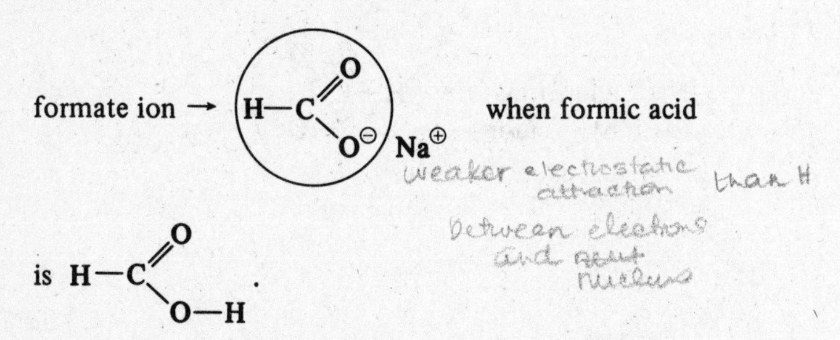

Write the structural formula for sodium acetate when

acetic acid is H—C(H)(H)—C(=O)—O—H. Circle the acetate ion.

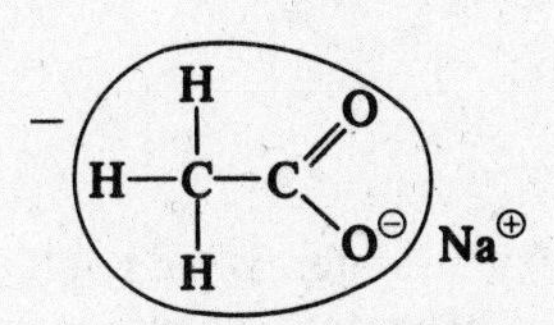

15) Even though sodium acetate is written

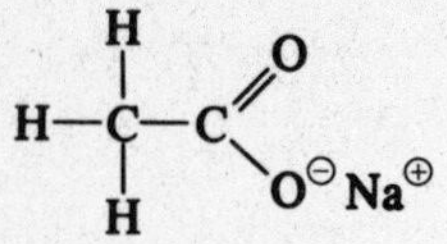

$$H-\overset{H}{\underset{H}{C}}-C\overset{\nearrow O}{\searrow O^{\ominus}\, Na^{\oplus}}$$

it is also a crystalline or an ionic compound. What is the negative ion (ionic group) in this organic salt (Na acetate)?

—Acetate ion $\left(H-\overset{H}{\underset{H}{C}}-C\overset{\nearrow O}{\searrow O^{\ominus}}\right)$

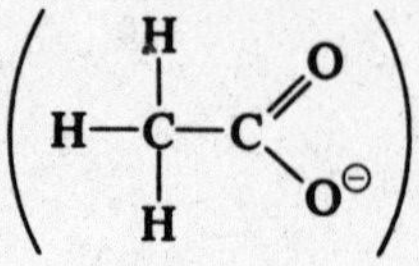

16) How many electrons are in the outer shell of the "lower oxygen" ($-O^{\ominus}$) of the acetate ion?

—8 $\left(-C\overset{\nearrow O}{\searrow :\ddot{\underset{..}{O}}:^{\ominus}}\right)$

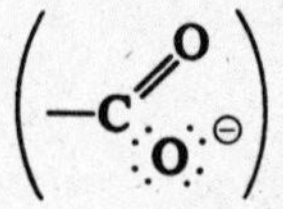

17) Organic salts are named in the same fashion as organic acids and aldehydes:

Aldehyde	Acid	Salt or Ion
Form/aldehyde	Form/ic Acid	Form/ate
Acetaldehyde	?	?

Fill in the (?) in the table. What ending is characteristic of an organic acid? What ending is characteristic of an organic salt or ion?

—Acetic acid, acetate
-ic (acid)
-ate (salt)

18) Label each of the following as salt or acid:

Lactic________
Fumarate________
Propionic________
Succinate________
Pyruvate________

—Lactic acid
Fumarate salt
Propionic acid
Succinate salt
Pyruvate salt

19) What ions are produced when sodium formate $\left(H-C\overset{\nearrow O}{\searrow O^{\ominus}\, Na^{\oplus}}\right)$ dissociates (forms ions)? Label the cation and the anion.

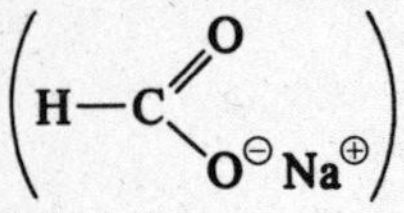

—$Na^{\oplus}$ and formate $^{\ominus}$ $\left(H-C\overset{\nearrow O}{\searrow O^{\ominus}}\right)$
(cation) (anion)

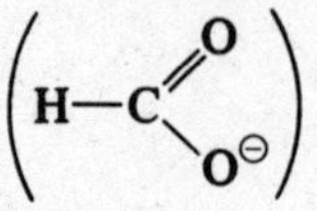

20) What ions are formed when $H-\overset{H}{\underset{H}{C}}-C\overset{\nearrow O}{\searrow O^{\ominus}\, Na^{\oplus}}$ dissociates?

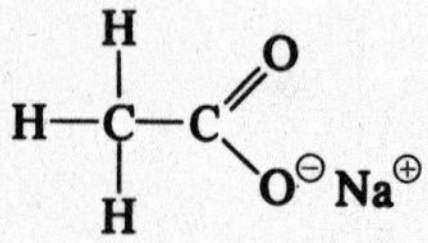

— $H-\overset{H}{\underset{H}{C}}-C\overset{\nearrow O}{\searrow O^{\ominus}}$ + $Na^{\oplus}$
(acetate)

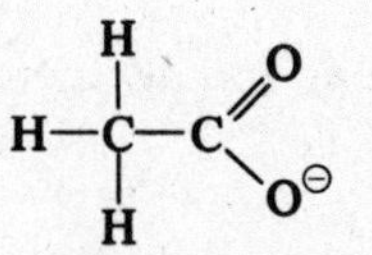

21) Most salt crystals are very soluble in water. When a salt dissolves, it dissociates into ions. What ions are produced when KCl dissociates in water?

—K^{+} and Cl^{-}

22) Dissociation can also be written as an equation:

$K^{\oplus}Cl^{\ominus} + H_2O \rightleftharpoons K^{\oplus}(aq) + Cl^{\ominus}(aq)$ (aq) indicates an aqueous medium.

The double arrow $\rightleftharpoons$ indicates that the dissociation is reversible (i.e., $K^{\oplus}$ and $Cl^{\ominus}$ can combine to form $K^{\oplus}Cl^{\ominus}$). Write the equation for the dissociation of $Na^{\oplus}F^{\ominus}$ in water.

–$Na^{\oplus}F^{\ominus} + H_2O \rightleftharpoons Na^{\oplus}(aq) + F^{\ominus}(aq)$

(23) Write the equation for the dissociation of $Ca^{++}Cl_2^{\ominus}$.

–$Ca^{++}Cl_2^{-} + H_2O \rightleftharpoons Ca^{++}(aq) + 2Cl^{\ominus}(aq)$

24) $Na^{\oplus}F^{\ominus} + H_2O \rightleftharpoons Na^{\oplus}(aq) + F^{\ominus}(aq)$

For each $Na^{\oplus}F^{\ominus}$, how many + charges are produced? –One

how many - charges are produced? –One

does the total + = the total -? –Yes

25) $Ca^{++}Cl_2^{\ominus} + H_2O \rightleftharpoons Ca^{++}(aq) + 2Cl^{\ominus}(aq)$

Total + charges produced?_____ –2

Total - charges produced?_____ –2

Are they equal?_____ –Yes

26) State a generalization about the relationship between the + and - charges on the ions that are produced when a salt dissociates.

–Total + = total - (or similar response)

(27) Would the following make sense? Explain you answer.

$Mg^{++}Cl_2^{\ominus} + H_2O \rightleftharpoons Mg^{++}(aq) + Cl^{\ominus}(aq)$

–No

Total + = 2
Total - = 1 They should be equal.

(28) All chemical bonding is due to what forces?

–Electrostatic attraction; attraction between + and - charges. (or similar response)

(29) In ionic compounds, what units are held together by electrostatic attraction?

–Ions (+ and -).

Take at least a 5-minute "break" before continuing.

UNIT FOUR POLARITY, HYDROGEN BONDS, SOLUBILITY, AND HYDROPHOBIC INTERACTIONS 4

CONCEPTS

Within an electrostatically neutral molecule (where $p^+ = e^-$), zones of partial ⊕ and ⊖ charges may result from the localized concentration or dispersal of electrons. For example, in a water molecule:

–The hydrogen atoms are not bonded symmetrically at opposite sides of the oxygen atom.

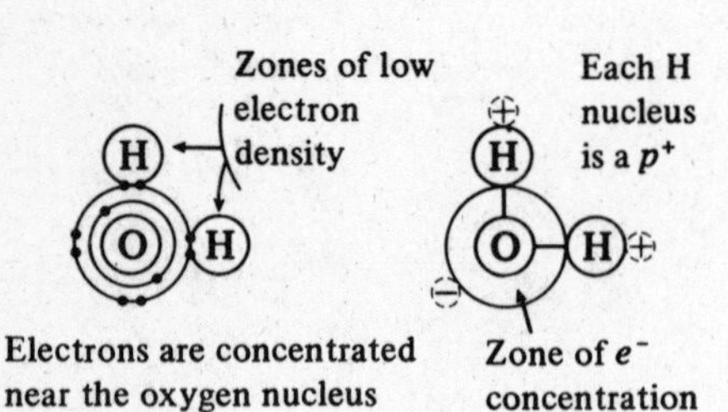

–The shared electrons are more strongly attracted to the oxygen nucleus (8+) than to the hydrogen nuclei (1+).

–The high concentration of electrons around the oxygen nucleus and the related deficiency of electrons around each hydrogen nucleus results in zones of partial electrostatic charges.

–Thus, electrostatic polarity is the presence of local zones of positive and negative charge within a molecule or group of atoms.

e^- distribution charges = ⊕ and ⊖ . The charges are less than those of full e^- and p^+ charges.

e^- transfer charges = ⊕ and ⊖ or + and –. Such ionic charges are equal to full e^- and p^+ charges due to electron transfer.

Hydrogen bond: the attraction between a ⊕ charged H atom and a ⊖ charged N or O atom.

Carbon-hydrogen bonds are not significantly polar in water solutions.

Hydrogen bonds

Dipolar (2)	Tripolar (3)	Polypolar (many)

Dipole-dipole bond: a general term for any bond that forms between two or more polar molecules by the attraction of a ⊕ atom to a ⊖ atom. A hydrogen bond is a type of dipole-dipole bond.

Solubility: the degree to which a substance can form H bonds or ion-dipole bonds with water molecules and thereby become dispersed among the water molecules.

Hydrophilic groups: polar or ionic groups that promote solubility in water.

Hydrophobic groups: nonpolar groups, typically hydrocarbons, that retard solubility in water.

Hydrophobic interactions: the clustering together of hydrophobic groups in water as a result of their insolubility in water (for example, a fat droplet in water). Hydrophobic interactions are extremely important in biological systems, especially the structure of proteins and biomembranes.

QUESTIONS AND PROBLEMS

SAMPLE RESPONSES

1) A hydrogen atom is composed of a single proton and a single electron. In the equation

$$H^{\oplus}Cl^{\ominus} + H_2O \rightleftharpoons H^{\oplus}(aq) + Cl^{\ominus}(aq),$$

what happens to the H's electron? What is a H^+ (hydrogen ion) in terms of protons and electrons?

–The electron is attached to the Cl^- (or similar answer).

–$H^{\oplus}$ = a proton (Actually the H^+ attracts water so strongly that it never occurs "naked" in aqueous medium, but is surrounded by ion-dipole bonded water molecules.)

2) The pair of electrons shared between some nuclei are unequally shared.

Example, in H_2O: H:O:H

The electrons are held more closely to the O than to the H.

Each H in this diagram is a single proton. What charge (⊕ or ⊖) would be associated with each H?

– ⊕

3) Detailed experimental analysis of water molecules reveals that the two H atoms tend to be located on one side of the O thus:

H H / O (not H—O—H) H :Ö:H (with dots below O)

Structural formula — Outer shell diagram

Remember the charges: electron (–) protons (+)

Label the areas of this water molecule in which the protons predominate and the area in which the electrons predominate.

H H / O

–

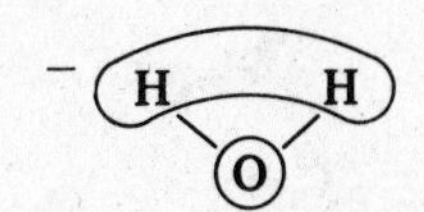

Protons predominate

Electrons predominate

4) Label one atom (end) of this water molecule ⊖ and *two* atoms (ends) ⊕.

H H / O

– ⊕H H⊕ / O⊖

5) Arrange these dipoles in a line with their ends touching.

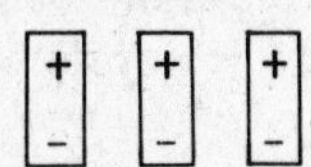

–

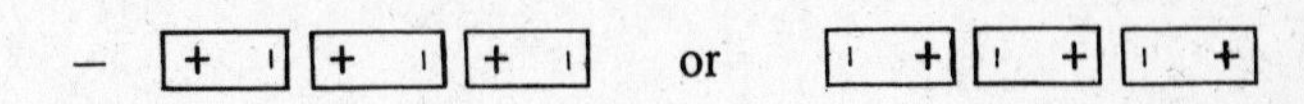

6) Draw two molecules of water with one H or one molecule attracted to the O of the other molecule to form a hydrogen bond.

– H H / O ··· H H / O (or other reasonable arrangement)

7) Each O can attract two H from different water molecules. Arrange three water molecules to show this formation of two hydrogen bonds.

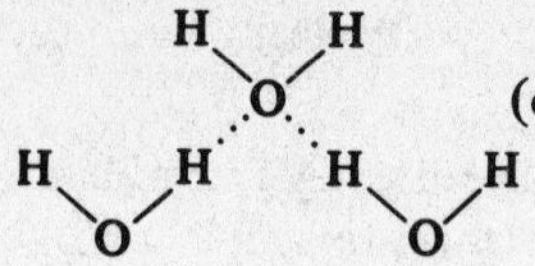

H H O ··· H H H H O O (or other reasonable arrangement)

8) The =O in a carboxyl group (—C(=O)OH) is also a ⊖ atom.

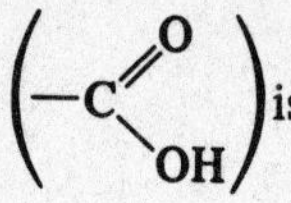

Diagram a H bond between the =O of a carboxyl group and a water molecule.

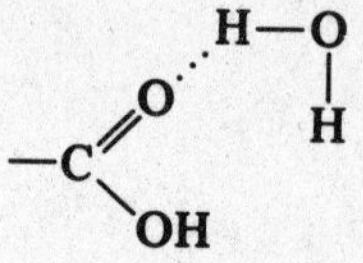

— —C(=O···H—O—H)OH

9) Diagram a H bond between the =O of a carboxyl group and an H in an amino group (—N(H)H)

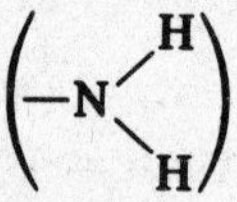

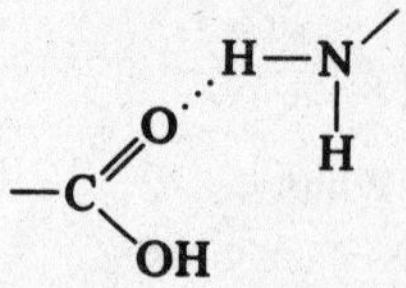

— —C(=O···H—N(H))OH (or similar diagram)

10) In the following structure, the right-hand N is a ⊖ atom. Diagram a H bond with a molecule of water.

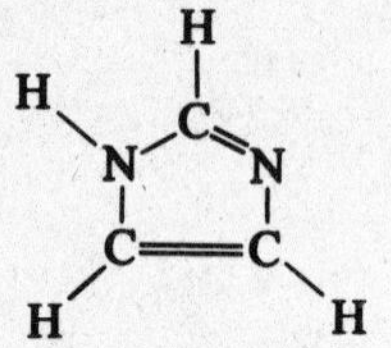

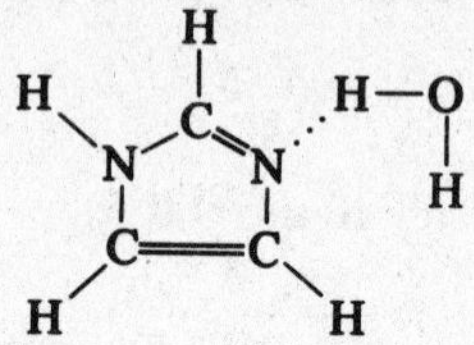

— imidazole ring with right-hand N···H—O—H

11) The following atoms are of particular importance in biological H bonding:

Dipolar H ⊕	Dipolar Atoms ⊖
H in —O—H as in hydroxyl and water	=O as in carboxyl groups
	O unbalanced bonded O as in water
	N unbalanced bonded N as in imidazole ring
H in —N(H)H amino groups	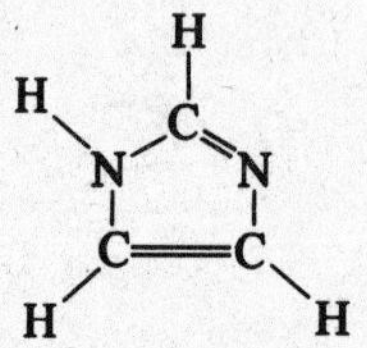

Define a H bond (you may use the above material).

—A H bond is the electrostatic attraction between a (+) dipolar H atom that is covalently bonded to an O or N atom and a (-) dipolar atom, such as an O or N. (or similar response)

12) Using the concepts you have learned, write a general definition for any dipole-dipole bond.

—A dipole-dipole bond is the attraction between the + end of one dipole and the - end of another dipole. (or similar response)

13) You have already learned that these groups are polar (charged).

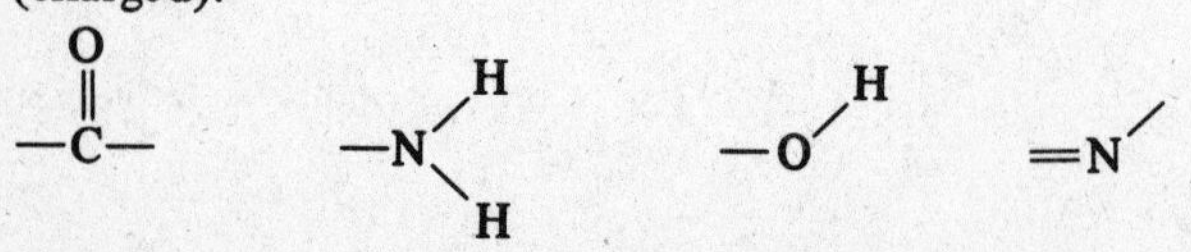

—C(=O)— —N(H)H —O—H =N—

Label the appropriate atoms with a charge ⊕ or ⊖.

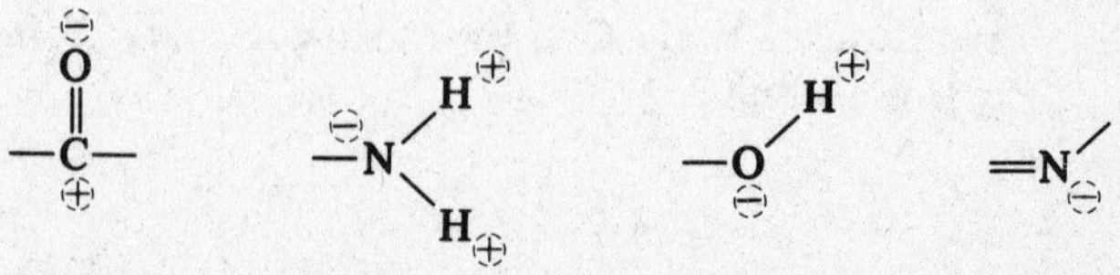

— —C⊕(=O⊖)— —N⊖(H⊕)H⊕ —O⊖—H⊕ =N⊖—

14) All —OH groups contain an unbalanced bonded O (—O—H) Rewrite the following molecules with OH groups to show this. Add the ⊕ and ⊖ to all the polar atoms.

```
   H                  O              O
   |                //               ||
H—C—OH          H—C             HO—P—OH
   |                \                |
   H                 OH              OH
```

—

```
   H     H⊕           O⊖            ⊖
   |    /           //     H⊕   ⊕H  O   H⊕
H—C—O⊖          H—C⊕     /       \ ⊕||  /
   |                \ O⊖            O⊖—P—O⊖
   H                                    |
                                        O⊖
                                         \
                                          H⊕
```

15) Diagram the attraction of two water molecules to this molecule of formaldehyde:

```
H
 \
  C=O
 /
H
```

—

```
H—O
   \
    H
     .
      O···H
     //     \
  H—C        O
     |      /
     H     H
```

(or similar diagram)

16) Diagram the attraction of water molecules to this molecule of methyl alcohol.

```
   H
   |
H—C—O
   |  |
   H  H
```

—

```
     O
    / \
   H H  H       H
   |    :       |
H—C—O···H—O
   |    \
   H     H
          .
           O
          / \
         H   H
```

(or similar diagram)

17) H that is covalently bonded to C is nonpolar in aqueous solution (relative to water it does not bear a charge)

```
   H  H
   |  |
H—C—C—
   |  |
   H  H
```

is thus a nonpolar group. It does not form H bonds. Is

```
   H  H  H
   |  |  |
H—C—C—C—
   |  |  |
   H  H  H
```

a polar or nonpolar group?

—Nonpolar

Will this group attract water molecules?

—No

18) Circle the nonpolar groups. Write the appropriate charge on the atoms of polar groups.

```
   H                 H  H  H
   |    H            |  |  |    O
H—C—O            H—C—C—C—C//
   |                 |  |  |    \   H
   H                 H  H  H     O/
```

—

```
 ( H )   H⊕
 ( | )  /
 (H—C—) O⊖
 ( | )
 ( H )
```

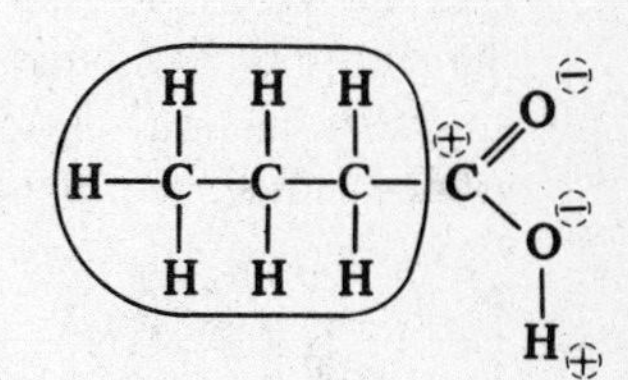

19) The solubility of a molecule in water is determined by its ability to attract water (form H bonds with water). Will polar groups increase or decrease solubility?

—Increase

Will nonpolar groups increase or decrease solubility?

—Decrease

20) On the following molecules, circle the groups that will increase the solubility of the molecule.

```
   H  O  H                H  H                 
   |  ‖  |                |  |    //O          
H—C—C—C—H             H—C—C—C               
   |     |                |  |    \OH          
   H     H                H  NH2               

   H  H  H
   |  |  |
H—C—C—C—H
   |  |  |
  OH OH OH
```

```
    H (O) H                 H  H    (O)
    |  ‖  |                 |  |   //
– H—C—C—C—H             H—C—C—C
    |     |                 |  |   \(OH)
    H     H                 H (NH2)

    H  H  H
    |  |  |
 H—C—C—C—H
    |  |  |
  (OH)(OH)(OH)
```

21) Ions are atoms or groups that bear a + or – charge. Would an ion be soluble in water? Explain your response briefly using a diagram.

–Yes

–Ions (+ or –) would attract water molecules. For example, a + ion would attract water thus:

```
   H       H
   |       |
 H—O       O—H
      (+)
 H—O       O—H
   |       |
   H       H
```

(22) What type of group is insoluble? What two types of groups or atoms are soluble?

–Insoluble: nonpolar groups (hydrocarbons)

–Soluble: ions and polar groups

23) Nonpolar groups are also called *hydrophobic groups* (water-fearing groups). Hydrocarbons are the most common hydrophobic groups in biological material. Circle and label the hydrophobic groups.

```
   H  H  H     H            H  H     H
   |  |  |    /             |  |     |
H—C—C—C—O              H—C—C—O—C—H
   |  |  |                  |  |     |
   H  H  H                  H  H     H
```

–

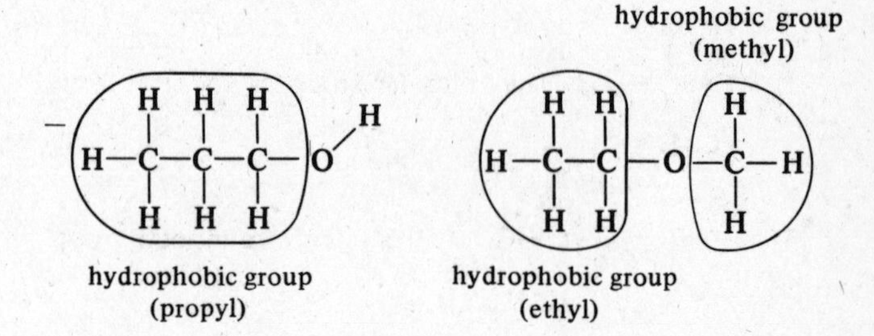

24) In an aqueous medium (i.e., in water), polar groups and ions will be surrounded by molecules of ______. Hydrophobic groups will tend to cluster together (e.g., "oil and water do no mix"). On these molecules, circle the groups that will be surrounded by water molecules and label them "H_2O," also circle the hydrophobic groups and connect these circles with a line to indicate clustering.

–Water

```
   H  H  H  H
   |  |  |  |
H—C—C—C—C—OH
   |  |  |  |
   H  H  H  H

H        H  H  H  H
 \       |  |  |  |
  N—C—C—C—C—H
 /       |  |  |  |
H        H  H  H  H
```

–

```
  [  H  H  H  H ]      H2O
  [  |  |  |  | ]       ↓
  [H—C—C—C—C ]——(OH)
  [  |  |  |  | ]
  [  H  H  H  H ]

   ( H   )   [ H  H  H  H    ]
   (  \  )   [ |  |  |  |    ]
   (   N )———[C—C—C—C—H ]
   (  /  )   [ |  |  |  |    ]
   ( H   )   [ H  H  H  H    ]
 H2O↗
```

25) The clustering together of hydrophobic groups in water is called hydrophobic interactions. The tendency toward hydrophobic interactions increases with the increase in length of the hydrocarbon chain. Which of the following would have the greatest tendency to form hydrophobic interactions?

–A

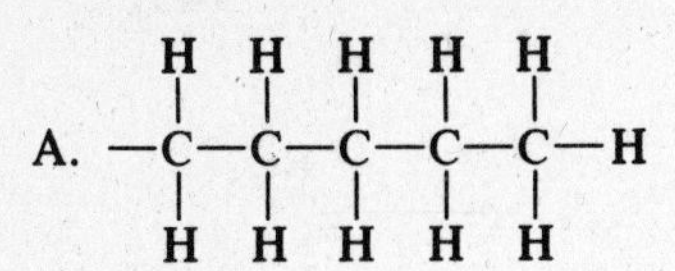

```
     H  H  H  H  H                 H
     |  |  |  |  |                 |
A. —C—C—C—C—C—H           C. —C—H
     |  |  |  |  |                 |
     H  H  H  H  H                 H

     H  H                  H  H  H  H
     |  |                  |  |  |  |
B. —C—C—H            D. —C—C—C—C—H
     |  |                  |  |  |  |
     H  H                  H  H  H  H
```

26) Which of the following groups would form hydrophobic interactions?

```
A. —OH          C. —NH2        E. —C(=O)O⊖ Na⊕

     H  H  H                          H  H
     |  |  |         D. —C(=O)OH      |  |
B. —C—C—C—H                       F. —C—C—H
     |  |  |                          |  |
     H  H  H                          H  H
```

–B and F

27) As you may have guessed, the preceding account of hydrophobic interactions is rather naive and oversimplified. A more adequate explanation is based on energetic and thermodynamic principles that we will not take time to develop. Thus, define hydrophobic interactions as well as you can at this time.

–Hydrophobic interactions are the clustering together of hydrophobic (or nonpolar or hydrocarbon) groups in an aqueous medium (in water). (or similar response)

(28) In a covalent bond, two atoms are held together by common attraction for a pair of electrons. In an ionic bond, 2 ions are held together by electrostatic attraction (+ and -). In a dipolar bond, 2 dipoles are held together by electrostatic attraction. In a hydrogen bond, a H atom is held to a ⊖ atom by electrostatic attraction. In you own words, what is the general definition of a bond?

–A bond is any force of electrostatic attraction that holds two or more atoms together. (or similar response)

(29) Label with the appropriate dipolar charges ⊕ or ⊖.

```
H       H                       ⊕H        H      O⊖
  \     |    O                     \  ⊖   |  ⊕  //
   N—C—C                             N—C—C        H⊕
  /     |    \                     /      |    \  /
H       H     O                 ⊕H        H     O
               \                               ⊖
                H
```

(30) Rediagram these molecules to illustrate hydrogen bonding between the =O and one H of the NH_2.

```
       O                   H H
      //           H       | |
H—C               N—C—C—H
      \           /  | |
       H         H   H H
```

—

```
                H          H H
                 \         | |
                  N—C—C—H
       O···H /   | |
      //              H H
H—C    ↖(three dots)
      \
       H
```

(31) On the diagram, circle the groups or portions of the molecule which will *retard* water solubility.

```
 H  H  H  H
 |  |  |  |     O
H—C—C—C—C—C //
 |  |  |  |    \
 H  |  H  |     OH
H—C—H   N
    |   /  \
    H  H    H
```

—

```
 H  H  H  H      (may be included)
 |  |  |  |     O
H—C—C—C—C—C //
 |  |  |  |    \
 H  |  H  |     OH
 H—C—H   N
    |    /  \
    H   H    H
```

(32) Define hydrophobic interactions.

—Hydrophobic interactions are the clustering together of hydrophobic groups in water. (or similar response)

Take at least a 5-minute break before continuing to the next unit.

UNIT FIVE EQUILIBRIUM — 5

CONCEPTS

Chemical reactions involve the formation or breaking of bonds between atoms and molecules.

Every chemical reaction has an equilibrium point at which the rate of the reaction to the right equals the rate to the left.

$CO_2 + H_2O \rightleftharpoons \underset{\text{Carbonic acid}}{H_2CO_3}$

By convention, reactants are usually written on the left and products are usually written on the right. However, which "side" of a reaction represents reactants and which represents products depends on the point of view and must be specified for any given reaction.

At equilibrium: rate to right = rate to left

If the concentrations of the reactants are increased, the rate of the reaction toward the products is increased: ($\overrightarrow{\rightleftarrows}$) and there is a net increase in products, and a new equilibrium will be reached.

If the concentrations of the reactants are decreased, the rate of the reaction toward the products is decreased: ($\underleftarrow{\rightleftarrows}$) and there is a net increase in reactants, and a new equilibrium will be reached.

QUESTIONS AND PROBLEMS

1) $A + B \rightleftharpoons C + D$
This reversible reaction is made up of a *forward reaction* and a *reverse reaction*. Write and label the forward and reverse reactions.

2) $A + B \rightleftharpoons C + D$
Consider the *reverse* reaction only.
What are the *products* of the reverse reaction? What are the *reactants* of the reverse reaction?

3) The *relative rates* of the forward and reverse reactions may be represented by adding an arrow above or below the reaction arrows. In the following, is the rate greatest for the forward or reverse reaction?

A. $CO_2 + H_2O \underleftarrow{\rightleftarrows} H_2CO_3$
B. $Q + R \overrightarrow{\rightleftarrows} S + T$
C. $ATP + H_2O \overrightarrow{\rightleftarrows} ADP + P$
D. $A + B \rightleftharpoons C + D$

4) This reaction is in *equilibrium*. At chemical equilibrium what is the relationship between the rates of the forward and reverse reactions?

$A + B \rightleftharpoons C + D$

5) Which of the following reactions are in equilibrium?

A. $A + B \rightleftharpoons C + D$
B. $A + B \overrightarrow{\rightleftarrows} C + D$
C. $ATP + H_2O \rightleftharpoons ADP + P$
D. $CO_2 + H_2O \underleftarrow{\rightleftarrows} H_2CO_3$
E. $ADP + P \rightleftharpoons ATP + H_2O$

SAMPLE RESPONSES

1) –Forward: $A + B \longrightarrow C + D$
–Reverse: $A + B \longleftarrow C + D$ (or $C + D \longrightarrow A + B$)
(All biochemical reactions are theoretically reversible.)

2) –Products: *A* and *B*
–Reactants: *C* and *D*

3) –A. Reverse
B. Forward
C. Forward
D. Equal

4) –The forward rate equals the reverse rate.

5) –A, C, and E (any order)

⑥ Define chemical equilibrium.

–Chemical equilibrium for a reversible reaction exists when the rate of the forward reaction equals the rate of the reverse reaction. (or similar response)

7) $A + B \overset{\longrightarrow}{\leftarrow} C + D$

For which reaction (forward or reverse) is the rate the greatest? Is the concentration of each of the following increasing or decreasing?

A________ B________ C________ D________

–Forward

–A decreasing B decreasing C increasing D increasing

8) LeChatelier's principle states that if a system at equilibrium is *stressed*, it will respond to *relieve* the stress and to obtain a new equilibrium.

$A + B \rightleftharpoons C + D$

This reaction starts at equilibrium. "A" is added to the system. What effect does this have on the concentration of "A"? This is a stress. Will an increase in the rate of the forward or reverse reaction relieve this stress? What happens to the concentration of "A" as the stress is relieved?

–Increased
–Forward reaction

–The concentration of A will decrease (the decrease in the concentration of A *is* the relief of the stress).

Pronounciation: LeChatelier
LOO (as in *look*) SHAH-TELL-YAY'

9) Indicate the response of this reaction to the stress by drawing a "relative rate arrow."

$A + B \rightleftharpoons C + D$ The concentration of D is increased (D is added).

–$A + B \underset{\longleftarrow}{\rightarrow} C + D$

⑩ State LeChatelier's principle.

–If a chemical system is stressed, it will react to relieve the stress and to attain a new equilibrium. (or similar response)

11) In the reaction $A + B \rightleftharpoons C + D$, if the concentration of A is increased, which reaction rate is increased? What are the products of the forward reaction? What are the reactants of the forward reaction?

–Forward
–Products–$C + D$
–Reactants–$A + B$

12) The *rate* of a reaction *depends on* the concentration of the *reactants only*.

$A + B \rightleftharpoons C + D$

The concentration of A is decreased. For which reaction is A a reactant? Thus, in which reaction will the rate be affected? Will this rate be decreased or increased? Which reaction rate is now greater? Indicate this (the relative rates) by adding a "relative rate arrow."

–Forward
–Forward
–Decreased
–Reverse
–$A + B \underset{\longleftarrow}{\rightarrow} C + D$

13) $A + B \rightleftharpoons C + D$

The concentration of C is decreased. Add a "relative rate arrow" to indicate the response to this stress.

–$A + B \overset{\longrightarrow}{\leftarrow} C + D$

14) Add a "relative rate arrow" for each of the following stresses:

Stress	Reaction: $A + B \rightleftharpoons C + D$
A. C increased	
B. B increased	
C. C decreased	
D. D decreased	
E. A and B increased	

–A. ←
B. →
C. →
D. →
E. →

15) $A + B \rightleftarrows C + D$

Is this reaction at equilibrium? Is the concentration of A increasing or decreasing? Is the concentration of C increasing or decreasing?

–No
–A is decreasing.
–C is increasing.

16) $A + B \rightleftharpoons C + D$

The concentration of A is increased. Add a "relative rate arrow." What is the effect of this increase in the concentration of A on the concentration of:

C________ D________ B________

–→ C increases D increases B decreases
(These changes continue until a new equilibrium is attained.)

(17) $Q + R \rightleftharpoons S + T$

The concentration of S is decreased. Add a "relative rate arrow." What is the effect on the concentration of:

T________ Q________ R________

–→ T increases Q decreases R decreases

18) In this reaction that occurs in mammalian muscle cells during exercise:

pyruvate + NADH $\rightleftharpoons$ lactate + $NAD^{\oplus}$

What effect would an increase in NADH concentration have on the reaction rate? What effect would an increase in NADH concentration have on lactate concentration?

–Rate of forward reaction increased (→).

–Lactate concentration increased.

(19) $CO_2 + H_2O \rightleftharpoons H_2CO_3 \rightleftharpoons H^+ + HCO_3^-$

This reaction occurs in the blood. In the lungs CO_2 (carbon dioxide) leaves the blood. What effect would this decrease in CO_2 concentration have on the H^+ concentration of the blood?

–H^+ will decrease.
($CO_2 + H_2O \leftleftarrows H_2CO_3 \leftleftarrows H^+ + HCO_3^-$)

Take at least a 5-minute break before continuing on to the next unit.

SELF-TEST UNITS 3, 4, & 5

(Criterion score 57 points: total of 61 points)

(3 points) 1) The atomic number of magnesium if 12.
In the magnesium ion what is the number of protons? Electrons?
What is the ionic charge?

(6 points) 2) Write the symbol for each of the following ions:

Na_____ Mg_____ K_____

Cl_____ Ca_____ F_____

(6 points) 3) Label each of the ions in #2 as a cation or anion.

(3 points) 4) Briefly, what is a salt?

(3 points) 5) Label each of the following as a salt or as an acid:

Acetate_______

Propionic______

Citrate________

(5 points) 6) Write the equation for the dissociation of the organic salt, sodium acetate ($CH_3COO^-Na^+(aq)$).

(3 points) 7) Define a hydrogen bond.

(5 points) 8) List the groups that commonly participate in H bonding in biological systems.

(3 points) 9) Define hydrophobic interactions.

(3 points) 10) Diagram a H bond between the right-hand N and an H of an amino group.

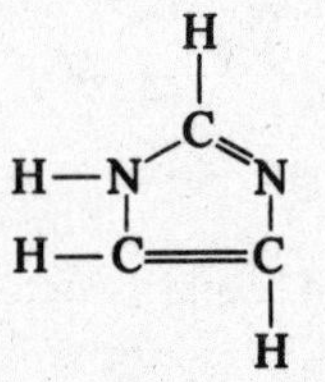

```
          H
          |
          C
        /   \\
  H—N         N
     |        |
  H—C ══════ C
              |
              H
```

(3 points) 11) What are the major factors that influence the solubility of a chemical substance in water?

(3 points) 12) Define chemical equilibrium.

(7 points) 13) Label the dipolar (partial) charges in this molecule.

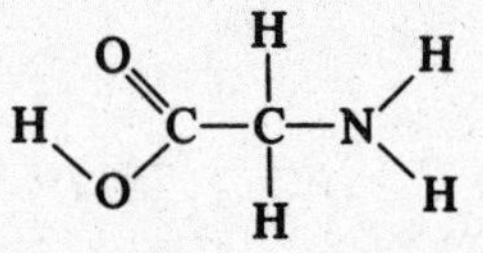

```
      O     H       H
       \\    |      /
  H     C — C — N
   \   /    |      \
    O       H       H
```

(3 points) 14) State LeChatelier's principle.

(5 points) 15) In the reaction:

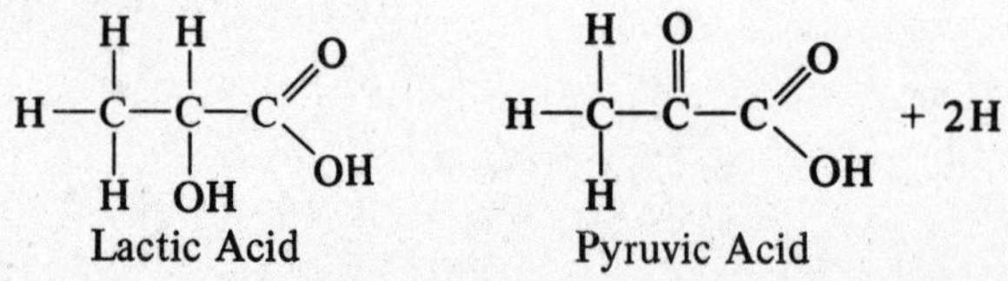

```
     H   H                        H   O
     |   |      O                 |   ||     O
  H—C — C — C//               H—C — C — C//        + 2H
     |   |      \                 |          \
     H   OH      OH               H           OH
    Lactic Acid                  Pyruvic Acid
```

If the concentration of pyruvic acid is decreased, what will be the effect on the concentration of lactic acid. Explain your response briefly.

(62 points total)

SELF-TEST KEY UNITS 3, 4, & 5

(Criterion score 57 points: total of 61 points)

(3 points) 1) Protons–12
Electrons–10
Net charge–+2 *(If you miss this, review Unit 3, Items 1–5.)*

(6 points) 2) Na^+ Mg^{++} K^+
Cl^- Ca^{++} F^- *(Review: Unit 3, Items 1–10)*

(6 points) 3) Cation Cation Cation
Anion Cation Anion *(Review: Unit 3, Item 10)*

(3 points) 4) A salt is an ionic compound (1) made up of metallic cations (1) and anions (1). *(Review: Unit 3, Items 11–13)*

(3 points) 5) Acetate–salt
Propionic–acid
Citrate–salt *(Review: Unit 3, Items 14–18)*

(5 points) 6) $CH_3COO^{\ominus}Na^{\oplus} + H_2O \rightleftharpoons CH_3COO^{\ominus}\ (aq) + Na^{\oplus}\ (aq)$

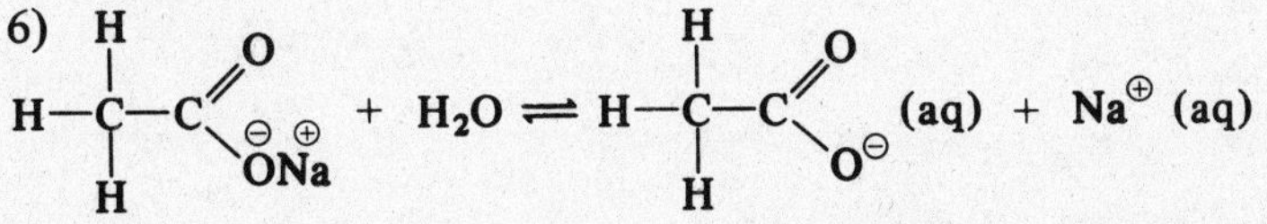

(Review: Unit 4, Items 1–9)

(3 points) 7) A hydrogen bond is the dipolar attraction (1) between a ⊕ H (1) and a ⊖ group (1). *(Review: Unit 4, Items 1–16)*

(5 points) 8) ⊕ H in OH ⊖ O in OH (1 point each)
H in NH_2 =O
=N– *(Review: Unit 4, Items 1–16)*

(3 points) 9) Hydrophobic interactions are the clustering together of hydrophobic groups (1), usually hydrocarbons (1), in water (1). *(Review: Unit 4, Items 17–27)*

(3 points) 10) [Imidazole ring structure: H–N, C–H, N···H–N– hydrogen bond, H–C=C–H] (three dots) *(Review: Unit 4, Items 6–10)*

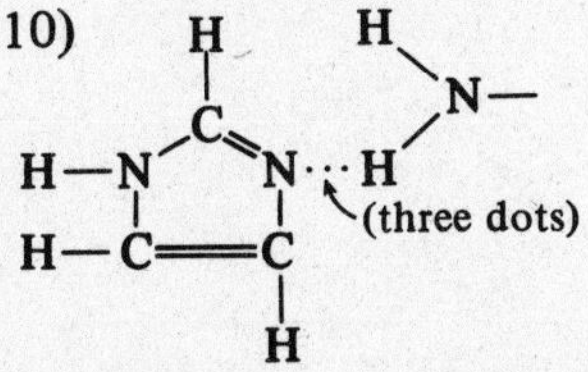

(3 points) 11) Solubility is influenced by the capacity to form H bonds (1) or ion-dipole bonds (1) with water molecules (1). *(Review: Unit 4, Items 19–22)*

(3 points) 12) For a reversible reaction at equilibrium the rate of the forward reaction equals the rate of the reverse reaction. *(Review: Unit 5, Items 1–6)*

(7 points) 13) [Glycine structure with partial charges: ⊖O=C⊕, ⊕H–O⊖, C–N⊖, ⊕H on N (two)] (1 point each)
(Review: Unit 4, Items 1–18)

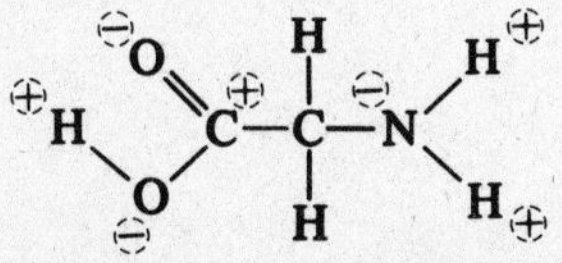

(3 points) 14) When a chemical system is stressed (1) it will react to relieve the stress (1) and to attain a new equilibrium (1). (*Review: Unit 5, Items 8–10*)

(5 points) 15) Lactic acid concentration decreases. Explanation: $\rightleftharpoons$; the rate of the reverse reaction is decreased. (or similar response) (5 points) (*Review: Unit 5, Items 8–19*)

Take at least a 5-minute break before continuing with the next unit.

UNIT SIX 6

MOLECULAR WEIGHT; MOLAR, PERCENT, MG PERCENT, AND PPM SOLUTIONS

CONCEPTS

Eventually most biologists have to get their hands dirty by mixing laboratory solutions. Here's how. . . .

The *molecular weight* of a molecule is the sum of the atomic weights of the constituent atoms. The unit of molecular weight is the dalton, which equals the weight of an H atom. For example:

	Atoms	*Atomic Weights*
CO_2	C	12
	O	16
	O	16
		M.W. = 44 daltons

	Atoms	*Atomic Weights*
Glucose: $C_6H_{12}O_6$	6 C	6 × 12 = 72
	12 H	12 × 1 = 12
	6 O	6 × 16 = 96
		M.W. = 180 daltons

The *gram molecular weight* of a molecule is the molecular weight expressed in grams. For example:

G.M.W. of CO_2 = 44 grams G.M.W. of glucose = 180 grams

A *mole* of a substance is one G.M.W.:

A mole of CO_2 = 44 grams of CO_2 A mole of glucose = 180 grams of glucose

For any gas: 1 mole = 22.4 liters at standard pressure and temperature (1 atmosphere and 23°C).

1000 ml 500 ml 100 ml

Solutions are usually made up in volumetric flasks (or graduated cylinders) that are calibrated to contain specific volumes of solutions, such as 100 ml, 500 ml, or 1 liter.

A 1 *molar solution* contains 1 gram molecular weight of a substance per liter of solution.

A 1 *percent solution* by weight contains 1 gram of solute per 100 grams of solution.

$$\text{General formula: } \% = \frac{\text{g solute}}{\text{g solution}}$$

A 1 percent solution by volume contains 1 ml (or 1 g) of solute per 100 ml of solution.

A 1 *parts per million* (ppm) solution contains 1 mg of solute per liter (1000 g) of solution.

$$\text{General formula: ppm} = \frac{\text{mg solute}}{\text{liters of solution}}$$

Milligram percent (mg%) solutions are used in clinical laboratories:

$$\text{mg\%} = \frac{\text{mg solute}}{\text{100 ml solution}}$$

Many chemists also use the "Atomic Mass Unit" as the unit of atomic and molecular weight.

1 atomic mass unit (a.m.u.) = 1 dalton

Table 3. Atomic Weights

Element	Atomic Number	Atomic Weight
Hydrogen	1	1.0
Helium	2	4.0
Lithium	3	6.9
Beryllium	4	9.0
Boron	5	10.8
Carbon	6	12.0
Nitrogen	7	14.0
Oxygen	8	16.0
Fluorine	9	19.0
Neon	10	20.1
Sodium	11	23.0
Magnesium	12	24.3
Aluminum	13	27.0
Silicon	14	28.1
Phosphorus	15	31.0
Sulfur	16	32.1
Chlorine	17	35.5
Argon	18	39.9
Potassium	19	39.1
Calcium	20	40.1

Prefixes for Small Scale

milli- = 10^{-3}
micro- = 10^{-6}
nano- = 10^{-9}
pico- = 10^{-12}

QUESTIONS AND PROBLEMS

SAMPLE RESPONSES

1) What are the *three* fundamental particles of which all atoms are constituted?

—Protons, electrons, neutrons (any order).

2) How many protons and electrons does the Al atom contain?

_____protons

_____electrons

—13 protons
13 electrons

3)

	Calcium
Atomic weight	40
Minus the number of protons	-20
Number of neutrons	20

How many neutrons are there in an F atom?

—10 neutrons (19 - 9)

4) What does the atomic weight of an atom equal (in terms of fundamental particles)?

—Number of protons + number of neutrons. (This is only a *rough* approximation, since natural substances may be composed of isotopic mixtures.)

5) Chlorine atoms contain: _____protons
_____electrons
_____neutrons

—17
17
18 and 19 (Actually in the natural mixture about half contain 18 and half 19.)

6) Calculate the molecular weight of NaOH.

—40 daltons
Na - 23
O - 16
H - 1
MW = 40 daltons

7) What is the gram molecular weight of NaOH?

—40 grams

8) Calculate the molecular weight and gram molecular weight of K_2SO_4.

—174.3 daltons
—174.3 grams

9) How many grams of NaOH would 1 liter of a 1 M solution contain?

—40 grams

10) How many grams of NaOH would 1 liter of a 0.3 M solution contain?

—12.0 grams (40.0 g × 0.3 M)

11) How many grams of NaOH would 400 ml of a 1 M solution contain? (1000 ml = 1 ℓ)

—16.0 grams ($40.0 \times \frac{400}{1000} = 40 \times 0.4$)

12) How many grams of NaOH would 600 ml of a 0.2 M solution contain?

—4.8 grams G.M.W. = 40 g
1 liter of 1 M = 40 g/ℓ
1 liter of 0.2 M = 8 g/ℓ
600 ml of 0.2 M = 4.8 g/600 ml

13) To mix a laboratory solution: The chemical to be dissolved is placed in the appropriate flask. Distilled water is then added until the flask is three-quarters full. The flask is swirled until all the chemical is dissolved. Then distilled water is added to the volume mark, and the solution is mixed thoroughly. Pour the solution into a *labeled** laboratory bottle.

List the steps for the preparation of 1 ℓ of a 1 M solution of NaOH.

—Calculate: 1 ℓ of 1 M NaOH = 40 g/ℓ.
Use a 1 liter volumetric flask.
Weigh out 40 g of NaOH; place it in the flask.
Add about 800 ml of distilled water.
Swirl the flask until all the NaOH is dissolved.
Add distilled water to the 1000 ml line.
Mix thoroughly.
Transfer to a labeled lab. bottle.
Clean and rinse the flask thoroughly—place it on a drying rack. (or similar response)

(14) Outline the procedures you would follow in the preparation of 500 ml of a 0.30 M solution of sodium bicarbonate ($NaHCO_3$).

—Calculate: 1 ℓ of 1 M $NaHCO_3$ = 84.0 g/ℓ.
1 ℓ of 0.30 M $NaHCO_3$ = 25.20 g/ℓ.
500 ml of 0.30 M $NaHCO_3$ = 12.60 g/ℓ.
Use a 500 ml volumetric flask.
Weigh out 12.60 g of $NaHCO_3$.
Add about 400 ml distilled water.
Swirl until all the $NaHCO_3$ is dissolved.
Add distilled water to the 500 ml mark.
Mix thoroughly.
Pour into a labeled lab. bottle.
Clean up thoroughly.

15) Some crystalline materials contain water of hydration in the crystal structure. This is included in the calculation of the molecular weight. What is the molecular weight of $MgCl_2 \cdot 6H_2O$?

—203 daltons
Calculation: Mg—24.3 × 1 = 24.3
Cl—35.5 × 2 = 71.0
H—1.0 × 12 = 12.0
O—16.0 × 6 = 96.0
203.3 daltons

*Most scientists are very fussy and idiosyncratic about how their laboratory solutions should be labeled. Therefore, if you mix real solutions, check with the *person in charge* to see how she or he wants it done.

(16) How many grams of $MgCl_2 \cdot 6H_2O$ would you use in preparing 1 liter of a 0.1 M solution?

–20.3 g/liter

(17) Outline the preparation of 100 ml of a 2% solution by weight of potassium chloride.

–Calculate: 2% = 2 g solute/100 g solution.
Weigh out 2 g of KCl.
Place the 2 g of KCl in the beaker.
Add 98 ml (98 g) distilled water using a graduated cylinder.
Mix thoroughly.
Transfer to a labeled lab. bottle.
Clean up. (or similar response)

(18) Outline the preparation of 500 ml of a 15% solution by volume of methyl alcohol (methanol).

–Obtain and label a 500 ml lab. bottle.
With a graduated cylinder, transfer 75 ml (15 × 5) of methanol to the bottle (for small volumes, volumetric pipettes are usually used for greater volumetric accuracy).
Similarly, add 425 ml of distilled water to the bottle.
Mix thoroughly.
Clean up.

(19) Outline the preparation of 200 ml of a 1% solution by volume of acetone.

–Obtain and label (about) a 200 ml lab. bottle.
With a volumetric pipette, transfer 2 ml of acetone to the bottle.
Add 198 ml distilled water.
Mix thoroughly.
Clean up.

→ (20) Outline the preparation of 1 liter of a 1% solution by volume of NaCl.

–Calculation: 1 ℓ 1% NaCl = 10 g/ℓ.
Weigh out 10.0 g of NaCl, place in a 1 ℓ volumetric flask.
Add about 800 ml distilled water.
Swirl until the NaCl is dissolved.
Add distilled water to the 1 ℓ mark.
Mix thoroughly, transfer to a labeled lab. bottle.
Clean up.

(21) Outline the preparation of 100 ml of a 5 ppm solution of IAA (a plant hormone) in water.

–Calculation: 5 ppm = 5 mg/ℓ
    100 ml 5 ppm = 0.5 mg/100 ml
Weigh out 0.5 mg of IAA.
Place it in a 100 ml volumetric flask.
Add 80 ml of distilled water, swirl until the IAA is dissolved.
Add distilled water to the 100 ml line.
Mix thoroughly.
Transfer to a labeled lab. bottle.
Clean up.

(22) Outline the preparation of 500 ml of a 100 mg% solution of glucose in water.

–Calculation: 100 mg% = 100 mg/100 ml.
    500 ml 100 mg% = 500 mg/500 ml.
Weigh out 500 mg of glucose.
Place it in a 500 ml volumetric flask.
Add about 400 ml of distilled water.
Swirl until dissolved.
Add distilled water to the 500 ml line.
Mix thoroughly and transfer to a labeled lab. bottle.
Clean up.

(23) Preparing laboratory solutions by dilution of "stock" solutions is common practice. Assume that you have a stock solution of 1 M HCl. How would you prepare 1 liter of a 0.10 M solution of HCl from this 1 M stock solution?

–Obtain and label a 1 ℓ laboratory bottle.
Transfer 100 ml of the 1 M HCl stock solution to the bottle with a graduated cylinder.
Add (carefully, because diluting acids produces heat) 900 ml distilled water with a graduated cylinder.
Mix thoroughly and clean up.

24) A simple equation for volumetric dilutions is:

$$V \times M = V' \times M'$$

volume × molarity = volume × molarity
stock solution desired solution

Use this formula to determine the volume of 1 M stock HCl you would use to make 80 ml of 0.5 M HCl.

–$V \times M = V' \times M'$
$V \times 1 = 80 \times 0.5$
$V = 40$ ml of 1 M stock solution

(25) Given a stock solution of 7 M HCl. How much of the stock would you use to make 200 ml of 0.3 M HCl?

–$V \times M = V' \times M'$
$V \times 7 = 200 \times 0.3$
$V = 60/7 = 8.56$ ml

26) Millimolar (mM) solutions are commonly used in biology. What is the molar concentration of a 5 mM solution?

–0.005 M (5/1000)

(27) Frog Ringer is a solution that simulates the salt content of the blood and tissue fluid. How much of each salt would you use to make 500 ml of a Ringer solution containing 120 mM of NaCl and 2 mM of KCl?

–NaCl G.M.W. = 58.5 daltons
1 M = 58.5 g/ℓ
120 mM = 58.5 × 0.120 = 7.02 g/ℓ
= 3.51 g/500 ml

KCl 1 M = 74.5 g/ℓ
2 mM = 74.5 × 0.002 = 0.149 g/ℓ
= 0.0745 g/500 ml

Take at least a 5-minute break before continuing with the next unit.

UNIT SEVEN pH, H^+ CONCENTRATION 7

CONCEPTS

The biological activities of many cellular constituents, particularly enzymes and other proteins, are greatly influenced by changes in pH.

p is a symbol for the −log

log N = power to which 10 must be raised to obtain N.

$$pH = -\log H^+ \text{ concentration} = \log \frac{1}{H^+ \text{ conc.}}$$

For example:

$$\log 100 = \log 10^2 = 2$$
$$\log 0.01 = \log 10^{-2} = -2$$
$$\log 0.00001 = \log 10^{-5} = -5$$
$$\log \frac{1}{0.0001} = \log \frac{1}{10^{-4}} = \log 10^4 = 4$$

Also:

$$\log (a \times b) = \log a + \log b$$
$$\log (a/b) = \log a - \log b$$

pH Scale

0 1 2 3 4 5 6 7 8 9 10 11 12 13 14

Very acidic — Neutral — Very basic

For distilled water: pH = 7.0

The pH of most cells and body fluids is about 7.0 to 7.5. This is called *physiological pH.*

Memorization Table (Approximate)
log 2 = 0.3
log 3 = 0.5
log 5 = 0.7
log 8 = 0.9

Two-Place Logarithms

Number	Logarithms
1	0.00
2	0.30
3	0.48
4	0.60
5	0.70
6	0.78
7	0.85
8	0.90
9	0.95

If you have not had experience with logarithms, don't panic; they are not as bad as they look at first glance. We'll introduce you to them gradually.

QUESTIONS AND PROBLEMS / SAMPLE RESPONSES

1) If acidity is high, basicity is_____.
If acidity is high, pH is_____.
High $[H^+]$ = ____ pH. ([ ] = concentration; $[H^+]$ = H^+ concentration).
Low $[H^+]$ = ____ pH.

—Low
Low
Low
High

2) What does the symbol "$[OH^-]$" mean?

—OH^- *concentration*

3) Complete this equation for the dissociation of HCl:

HCl (aq) ⟶

—HCl (aq) ⟶ H^+ (aq) + Cl^- (aq)

4) HCl dissociates completely in dilute aqueous solutions. Which of the following solutions is the most *acid*?
0.1 M HCl
0.01 M HCl
0.02 M HCl
0.005 M HCl

–0.1 M (this has the highest H^+ concentration)

Which is the most basic? Which has the lowest pH? Which has the highest pH?

0.005 M 0.1 M 0.005 M

5) Which of the following is most *basic*?
0.6 M HCl
0.02 M HCl
0.004 M HCl
0.001 M HCl

–0.001 M HCl

Which has the lowest pH?

–0.6 M HCl

6) If you are facile with logarithms you may skip to item 24. Try the following quiz to ascertain your facility (use the tables).

A. 10^{-3} = ________

B. log 1000 = ________

C. log 0.01 = ________

D. log 30 = ________

E. antilog 2.7 = ________

F. antilog -3.3 = ________

G. $\log (a \times b)$ = ________

H. $\log \frac{a}{b}$ = ________

I. $\log \frac{1}{x}$ = ________

J. from memory: log 2 = _____
log 3 = _____
log 5 = _____
log 8 = _____

–A. 0.001 or $\frac{1}{1000}$

B. 3

C. -2

D. 1.48 (or 1.5)

E. 500

F. 0.0005

G. $\log a + \log b$

H. $\log a - \log b$

I. $-\log x$

J. log 2 = 0.3
log 3 = 0.5
log 5 = 0.7
log 8 = 0.9

A score of 7 or more indicates reasonable facility.

7) Powers: 10^2 = 10 to the second power (or 10 squared)
10^3 = 10 to the third power (or 10 cubed)
10^4 = 10 to the fourth power

$10^0 = 10 \times \frac{1}{10} = 1$

$10^1 = 10$
$10^2 = 10 \times 10 = 100$
$10^3 = 10 \times 10 \times 10 = 1000$
10^4 = _____
10^5 = _____
10^8 = _____

–10^4 = 10,000 (4 zeros)
10^5 = 100,000 (5 zeros)
10^8 = 100,000,000

8) The logarithm (log) of 100 is 2 ($10^2 = 100$) (10 raised to the 2nd power)
log 1000 = 3 ($10^3 = 1000$) (10 raised to the 3rd power)
log 100,000 = 5 ($10^5 = 100,000$) (10 raised to the 5th power)
What is the logarithm of a number?

–The log of a number is the power to which 10 must be raised to give the number. (or similar response)

9) log 100 = _____
log 10,000 = _____
log 10,000,000 = _____

–2
4
7

10) $10^0 = 10 \times \frac{1}{10} = 1$

$10^{-1} = 10 \times \frac{1}{10} \times \frac{1}{10} = 0.1$ (decimal point moved over 1 place)

$10^{-2} = \frac{1}{10} \times \frac{1}{10} = 0.01$ (decimal point moved over 2 places)

$10^{-3} = \frac{1}{10} \times \frac{1}{10} \times \frac{1}{10} = 0.001$ (decimal point moved over 3 places)

log 0.01 = -2
log 0.001 = _____
log 0.00001 = _____

– -3
-5

11) log ($a \times b$) = log a + log b
log 20 = log (2 × 10) = __________
log 10 = 1
But, what is the log of 20? Then look up log 2 in Panel D, and solve the problem.

–log 2 + log 10

–1.3 (log 2 + log 10 = 0.3 + 1.0)

12) What is log 500?

–log 500 = 2.70 (Calculation: log 500 = log 5 + log 100
= 0.70 + 2
= 2.70)

13) What is log 0.3? (Solve as you did the last item.)

–log 0.3 = -0.5 [Calculation: log 0.3 = log 3 + log 0.1
= 0.5 + (-1)
= 0.5 - 1
= -0.5]

14) What is log 0.0008?

–log 0.0008 = -3.1 [Calculation: log 0.0008 = log 8 + log 0.0001
= 0.9 + (-4)
= -3.1]

15) log (?) = 2 (*anti*log of 2 = ?)
What is the antilog of 5?
log (?) = -3

–100
100,000
0.001

16) log (?) = 2.78

–600 [Calculation: log (?) = 2.78
= 2 + 0.78
? = 100 × 6
? = 600

What is the antilog of 3.85?

–7000
log ? = 3.85 = 3 + 0.85
? = 1000 × 7]

17) What is the antilog of -2.3?

—0.005 (Calculation: $\log ? = -2.3$
$= -3 + 0.7$
$? = 10^{-3} \times 5$
$? = 0.005$)

18) $\log (?) = -4.05$

—0.00009 [Calculation: $\log (?) = -4.05$
$= -5 + 0.95$
$? = 10^{-5} \times 9$
$? = 0.00009$]

(19) *Without* consulting Panel H:

log 8 = _____

log 3 = _____

log 2 = _____

log 5 = _____

—log 8 = 0.9

log 3 = 0.5

log 2 = 0.3

log 5 = 0.7 (Memorize these!)

20) Also: $\log \frac{a}{b} = \log a - \log b$

What is $\log \frac{1}{2}$?

—$\log \frac{1}{2} = \log 1 - \log 2 = 0 - 0.3$
$= -0.3$

21) What is $\log \frac{1}{a}$?

—$\log \frac{1}{a} = -\log a$ (Calculation: $\log \frac{1}{a} = \log 1 - \log a$
$= 0 - \log a$
$= -\log a$)

22) What is $\log \frac{1}{[H^+]}$?

—$\log \frac{1}{[H^+]} = -\log [H^+]$

23) Repeat the logarithm quiz (item 6).

24) The symbol p = the negative logarithm
= -log.
Thus, what is the pH ($p[H^+]$)?

— $-\log [H^+]$ (The negative logarithm of the hydrogen ion concentration.)

(25) Define pH.

—$pH = -\log [H^+]$ $\left(\text{or} = \log \frac{1}{[H^+]}\right)$

26) What is the pH of a 0.01 M HCl solution?

—pH = 2 $\quad pH = -\log [H^+]$
$= -\log 0.01 = -\log 10^{-2}$
$= -(-2) = 2$

(27) What is the pH of a 0.005 M HCl solution?

—pH = 2.3 [Calculation: $pH = \log \frac{1}{[H^+]} = \log \frac{1}{0.005}$
$= \log \frac{1}{5 \times 10^{-3}}$
$= \log \left(\frac{1}{5} \times \frac{1}{10^{-3}}\right)$
$= \log (0.2 \times 10^{3})$
$= \log (2 \times 10^{2})$
$= \log 2 + \log 10^{2}$
$= 0.3 + 2 = 2.3$]

(28) What is the pH of a 0.000005 M solution of H_2SO_4? (*Note:* each H_2SO_4 dissociates to produce $2H^+$.)

–pH = 5 [Calculation: $pH = \log \frac{1}{[H^+]}$
0.000005 M H_2SO_4
= 0.00001 M H^+

$$= \log \frac{1}{0.00001}$$
$$= \log \frac{1}{10^{-5}}$$
$$= \log 10^5$$
$$= 5]$$

29) In water solutions, $H_2O \rightleftharpoons H^+ (aq) + OH^- (aq)$. If $[H^+]$ is increased, what effect will this have on $[OH^-]$?

–$[OH^-]$ will decrease ($H_2O \rightleftharpoons H^+ + OH^-$)
(*Review: Unit 5*)

30) This relationship between $[H^+]$ and $[OH^-]$ can be expressed by the equation: pH + pOH = 14. If the pOH is 3.0, what is pH?

–pH = 11

31) Complete the equation for the dissociation of NaOH:

NaOH(aq) ⟶

–$NaOH(aq) \longrightarrow Na^+ (aq) + OH^- (aq)$

NaOH dissociates completely in water. What is the pOH of a 0.01 M solution of NaOH?

pOH = 2

32) What is the *pH* of a 0.1 M solution of NaOH?

–pH = 13 (pOH = 1; pH + pOH = 14;
pH = 14 - 1 = 13)

(33) How would you prepare 1 liter of a solution of NaOH that would have a pH of 12.7?

–One-liter volumetric flask. Use 2.0 g NaOH. Add 800 ml distilled H_2O. Swirl until dissolved. Add distilled H_2O to mark, mix, and transfer to a labeled bottle.

Sample calculation:
$$pH + pOH = 14$$
$$12.7 + pOH = 14$$
$$pOH = 1.3$$

$$pOH = \log \frac{1}{[OH^-]} = 1.3$$
$$= 0.3 + 1$$
$$= \log 2 + \log 10$$
$$= \log (2 \times 10) = \log (0.2 \times 10^2)$$
$$= \log \frac{1}{5} \times \frac{1}{10^{-2}} = \log \frac{1}{5 \times 10^{-2}}$$
$$\log \frac{1}{[OH^-]} = \log \frac{1}{5 \times 10^{-2}}$$
$$[OH^-] = 5 \times 10^{-2}$$

OH^- = 0.05M
M.W. = 40.0
0.05 M = 2 g/liter

34) $H_2O (d) \rightleftharpoons H^+ (aq) + OH^- (aq)$. What is the pH of distilled H_2O?

–pH = 7 [Calculation: $[H^+] = [OH^-]$
pH = pOH
pH + pOH = 14
2pH = 14
pH = 7]

Take a 5-minute break before going on to the next unit.

UNIT EIGHT — WEAK ACIDS, DISSOCIATION CONSTANTS, AND BUFFERS — 8

CONCEPTS

Since the biological activities of enzymes and other proteins are so sensitive to pH, biologists attempt to regulate the pH of experimental solutions with buffer systems.

A buffer system resists a change in pH. It consists of an aqueous solution of a weak acid (or base) and the salt of the acid.

For this equation for the dissociation of a weak acid: $HA(aq) \rightleftharpoons H^+(aq) + A^-(aq)$

the dissociation constant is expressed as: $K = \dfrac{[H^+]\ [A^-]}{[HA]}$

Weak Acids: Dissociation Constants and pK Values

Weak Acids	Dissoc. Const.	pK
$H_3PO_4(-H_2PO_4^-)$	7.52×10^{-3}	2.12
Lactic	8.4×10^{-4}	3.08
Formic	1.78×10^{-4}	3.75
$H_2CO_3(-HCO_3^-)$	1.70×10^{-4}	3.77
Succinic (1)	6.86×10^{-5}	4.16
Acetic	1.76×10^{-5}	4.75
Butyric	1.54×10^{-5}	4.81
Propionic	1.34×10^{-5}	4.87
Succinic (2)	2.47×10^{-6}	5.61
$H_2PO_4^-(-HPO_4^=)$	6.23×10^{-8}	7.21
"Tris" (trishy-droxy methyl-aminomethane)	5.0×10^{-9}	8.3

The Henderson-Hasselbalch Equation:

$$pH = pK + \log \frac{[A^-]}{[HA]}$$

*Most values were obtained from *Handbook of Chemistry and Physics* (1975), Chemical Rubber Co., Cleveland, Ohio.

QUESTIONS AND PROBLEMS

1) Write the formula for the dissociation constant for acetic acid (CH_3COOH or simply HAc).

$HAc \rightleftharpoons H^+ + Ac^-$

—$K = \dfrac{(H^+)(Ac^-)}{(HAc)}$

2) If one molecule of acetic acid dissociates, how many H^+ and how many Ac^- will be produced?

—One H^+ and one Ac^-

3) K for acetic acid = 1.76×10^{-5}. For each (one) molecule of acetic acid that dissociates, how many molecules remain undissociated? [That is, in the equation let $(H^+) = 1$ and $(Ac^-) = 1$, and then solve the equation.]

—$1.76 \times 10^{-5} = \dfrac{1 \times 1}{(HAc)} = \dfrac{1}{(HAc)}$

$(HAc) = \dfrac{1}{1.76 \times 10^{-5}} = \dfrac{1}{1.76} \times 10^5 = 0.568 \times 10^5$

$(HAc) = 56{,}800$

4) Express the relationship between the dissociated molecules of acetic and the undissociated molecules as a ratio. This is characteristic of a weak acid.

—1/56,800 (dissociated/undissociated)

5) Hydrochloric (HCl), sulfuric (H_2SO_4), and nitric (HNO_3) acids dissociate completely in water. Are they strong acids or weak acids?

—Strong acids

6) Using the table of dissociation constants, which of the following is the weakest acid: formic, acetic, or propionic?

–Propionic acid

7) What is the pK?

–- log K or log $\frac{1}{K}$

As the K value decreases the strength of an acid __?__.

–Decreases

As the pK value decreases the strength of an acid __?__.

–Increases

8) Write the equation for the dissociation constant for a weak acid. Multiply both sides of the equation by (HA). Cancel terms whenever this is possible.

$$\text{–(HA) K} = \frac{(H^+)(A^-)\,(HA)}{(HA)}$$

9) Next, divide both sides of the equation by (A^-).

$$–\frac{(HA)\,K}{(A^-)} = \frac{(H^+)(A^-)}{(A^-)}$$

10) Next, express both sides of the equation as reciprocals (1/*x*).

$$–\frac{1}{(H^+)} = \frac{1}{K} \times \frac{(A^-)}{(HA)}$$

11) And now, take the log of both sides of the equation.

$$–\log \frac{1}{(H^+)} = \log \frac{1}{K} + \log \frac{(A^-)}{(HA)}$$

12) And then, express $\log \frac{1}{X}$ as pX.

$$\text{–pH} = \text{pK} + \log \frac{(A^-)}{(HA)}$$

13) Congratulations! You have just derived the Henderson-Hasselbalch equation. It is very useful in predicting the behavior of buffers and weak acids.

Use $\text{pH} = \text{pK} + \log \frac{(A^-)}{(HA)}$ to calculate the proportion of acetic acid molecules that will be dissociated at pH = 5.06. (Subsitute known values and solve the equation.)

$$–5.06 = 4.76 + \log \frac{(A^-)}{(HA)}$$

$$0.30 = \log \frac{(A^-)}{(HA)}$$

(antilog of 0.30 = 2; from memory or the table in Unit 6)

$$2 = \frac{(A^-)}{(HA)} = \frac{\text{dissociated}}{\text{undissociated}} = \frac{2}{1}$$

(14) Similarly, calculate the proportion of $H_2PO_4^-$ ions that are dissociated at pH = 7.9.

$$H_2PO_4^- \rightleftharpoons H^+ + HPO_4^=$$

$$–7.9 = 7.2 + \log \frac{(HPO_4^=)}{(H_2PO_4^-)}$$

$$0.7 = \log \frac{(HPO_4^=)}{(H_2PO_4^-)}$$

$$5 = \frac{(HPO_4^=)}{(H_2PO_4^-)} = \frac{\text{dissociated}}{\text{undissociated}} = \frac{5}{1}$$

(15) One more: Calculate the proportion of $H_2PO_4^-$ ions that are dissociated at pH = 5.2. (*Hint:* Remember that

$-\log X = \log 1/X$ or $-\log a/b = \log b/a$.)

$$–5.2 = 7.2 \log \frac{(HPO_4^=)}{(H_2PO_4^-)}$$

$$-2.0 = \log \frac{(HPO_4^=)}{(H_2PO_4^-)} \quad \text{(Now, multiply both sides by -1.)}$$

$$2.0 = -\log \frac{(HPO_4^=)}{(H_2PO_4^-)} = \log \frac{(H_2PO_4^-)}{(HPO_4^=)}$$

$$10^2 = 100 = \frac{(H_2PO_4^-)}{(HPO_4^=)} = \frac{\text{undissociated}}{\text{dissociated}} = \frac{100}{1}$$

16) HA $\rightleftharpoons$ $H^+ + A^-$

If base (OH^-) is added, with which component of the above reaction would the OH^- react?

–H^+ ($H^+ + OH^- \rightleftharpoons H_2O$)

17) HA $\rightleftharpoons$ $H^+ + A^-$

If acid (H^+) is added, with which component of the above reaction would the (H^+) react?

–A^-(HA $\rightleftharpoons$ $H^+ + A^-$; as H^+ is increased).

18) Weak acid $\rightleftharpoons$ H^+ + salt anion$^-$

This is a buffer system–it buffers or resists a change in pH. As pH is increased does H^+ or OH^- increase?

–OH^- increases.

Which component of the buffer system resists this change?

–H^+ (reacts with the OH^-).

As pH is decreased does H^+ or OH^- increase?

–H^+ increases.

Which component of the buffer system resists this change?

–Salt anion$^-$ (reacts with the H^+).

19) What are the components of a buffer system?

–Weak acid + salt anion (of the weak acid)

How does a buffer system function?

–The buffer system resists a change in pH.
(H^+ reacts with OH^-.)
(Salt anion reacts with H^+.)

20) A weak acid dissociates only slightly. Would it produce much salt anion?

–No

The buffer capacity (salt anion) against (H^+) can be increased by adding a salt (e.g., sodium acetate) that dissociates completely in water. In the buffer system:

$$\text{HAc} \rightleftharpoons H^+ + Ac^- + Na^+ \rightleftharpoons \text{NaAc}$$

Does the H^+ come predominantly from the weak acid or from the salt?

–H^+ comes from the weak acid.

Does the Ac^- come predominantly from the weak acid or from the salt?

–Ac^- comes from the salt.

21) If a buffer system (composed of a weak acid and its salt) is to resist changes in pH by acids and bases with equal effectiveness, what should be the relationship between the acid and salt concentrations?

–They should be equal.

22) Of what components would an efficient buffer system be composed?

–A weak acid and its salt in equal concentrations.

23) The relationship between the pH and the relative concentrations of weak acid and salt can be expressed by the Henderson-Hasselbalch equation for a buffer system:

$$\text{pH} = \text{pK} + \log \frac{(\text{Salt})}{(\text{Acid})}$$

where
pK = the – log of the dissociation constant of the weak acid.*

What does the symbol "(acid)" mean?

–(Acid) = the concentration of the weak acid

What does the symbol "(salt)" mean?

–(Salt) = the concentration of the salt of the weak acid.

*The dissociation constant for any weak acid can be found in a *Handbook of Chemistry* or a *Handbook of Chemistry and Physics.*

24) In the Henderson-Hasselbalch equation for a buffer system, what does pH equal if (acid) = (salt)?

–pH = pK $\left(pH = pK + \log \frac{(salt)}{(acid)}\right.$

$pH = pK + \log 1 \qquad (acid) = (salt)$

$pH = pK + 0$

$\left. pH = pK \right) \qquad \frac{(salt)}{(acid)} = 1$

25) What is the relationship between (acid) and (salt) when the buffer capacity is greatest?

–(acid) = (salt)

Under these conditions, what is the relationship between pH and pK?

–pH = pK

For the greatest buffer capacity, the pH should = ____.

–pK

26) Write the Henderson-Hasselbalch equation for a buffer system.

–$pH = pK + \log \frac{(salt)}{(acid)}$

27) What is the name of this equation?

$$pH = pK + \log \frac{(salt)}{(acid)}$$

–Henderson-Hasselbalch equation for a buffer system.

(28) For the maximal buffer capacity at pH = 4.7, which of the following weak acids would you choose?

	pK
Phosphoric	2.1
Acetic	4.8
Carbonic	6.4

–Acetic

Explain you choice.

–pK is closest to desired pH,
(4.8 as compared to 4.7)
(pK = pH for maximal buffer capacity)

(29) If you wish to make up a laboratory solution to suspend isolated mitochondria at a physiological pH of 7.5, which of the buffers in the table would you choose?

–$H_2PO_4^- + HPO_4^=$ (pK = 7.2)

(30) To obtain a pH of 7.5, what proportions of NaH_2PO_4 (monobasic) and Na_2HPO_4 (dibasic) would you use?

–$pH = pK + \log \frac{(salt)}{(acid)} \qquad$ salt = Na_2HPO_4, acid = NaH_2PO_4

$$7.5 = 7.2 + \log \frac{Na_2HPO_4}{NaH_2PO_4}$$

$$0.3 = \log \frac{Na_2HPO_4}{NaH_2PO_4}$$

$$2 = \frac{Na_2HPO_4}{NaH_2PO_4} = \frac{2}{1}$$

(31) Given stock solutions of: 1 M NaCl
100 mM Na_2HPO_4
100 mM NaH_2PO_4

What amounts of each stock solution and of distilled water would you use to make up 1 liter of an experimental solution of 120 mM NaCl and a total of 3 mM sodium phosphate at pH = 7.5?

–$Na_2HPO_4/NaH_2PO_4 = 2/1$ (from question 30)
$V \times M = V' \times M'$
$V \times 100 \text{ mM} = 1000 \text{ ml} \times 3 \text{ mM}$
$$V = \frac{1000 \times 3}{100} = 10 \times 3 = 30 \text{ ml total Na phosphate}$$
Thus, use 20 ml Na_2HPO_4 and 10 ml NaH_2PO_4.
–For NaCl: $V \times M = V' \times M'$
$V \times 1000 = 1000 \times 120$
$$V = \frac{1000 \times 120}{1000} = 120 \text{ ml}$$
–Distilled water = 1000 - 120 - 3 = 877 ml
(Then check it with a pH meter, just to be sure!)

(32) At physiological pH, will most organic acids be associated or dissociated?

–Dissociated

33) Phosphate is an excellent buffer for physiological systems, but it has one drawback. If Ca^{++} is present, calcium phosphate may be formed, and calcium phosphate is only moderately soluble in water. At pH = 7.2, a precipitate of calcium phosphate will usually appear in solutions which contain more than about 3 mM of total phosphate ions and 3 mM Ca^{++}. Hence, what is the approximate upper limit for Ca^{++} and total phosphate ion concentrations in experimental solutions?

–About 3 mM maximum for both Ca^{++} and total phosphate ions.

(34) Explain why it is difficult to use high concentrations of phosphate as a buffer in physiological and biochemical systems.

–Physiological pH = about 7.2.
Calcium is present in most physiological systems.
Calcium phosphate is insoluble if the phosphate concentration exceeds about 3 mM.
Hence phosphate and Ca^{++} concentrations must be low and the buffer capacity will be low.

Take a 5-minute break before continuing.

SELF-TEST UNITS 6, 7, & 8

(Criterion score 65 points: total of 72 points)

(4 points) 1) Calculate the molecular weight of pyruvic acid ($CH_3COCOOH$).

(1 point) 2) What is the gram molecular weight of pyruvic acid ($C_3H_4O_3$)?

(14 points) 3) List the steps for the preparation of 500 ml of a 0.2 M solution of pyruvic acid.

(3 points) 4) How much pyruvic acid would you use to make 200 ml of a 3% solution by weight?

(3 points) 5) How much methanol would you use to make 800 ml of a 20% solution by volume?
How much distilled water would you add?

(3 points) 6) How much estrogen would you use to make up 2 liters of a 2 ppm solution?
How much distilled water would you use?

(3 points) 7) How much NaCl would you use in the preparation of 5 liters of a 950 mg% solution?

(4 points) 8) You have a stock solution of 7 M HNO_3 (nitric acid). How much stock solution would you use to prepare 500 ml of a 0.3 M solution?
How much distilled water would you use?

(1 point) 9) Express 0.0008 in mM.

(4 points) 10) Complete this table: log 2 =
log 3 =
log 5 =
log 8 =

(3 points) 11) Define pH.

(5 points) 12) Calculate the pH of a 0.008 M HCl solution.

(5 points) 13) What weight of NaOH would you use to prepare 1 ℓ of a solution of pH = 12?

(5 points) 14) The pK for acetic acid is 4.75. What proportion of the molecules are dissociated at pH = 5.25?

(5 points) 15) To obtain maximum buffer capacity at pH = 5.5, which of the following acids would you choose?

Acid	pK
Formic	3.75
Butyric	4.81
Succinic (2)	5.61

Explain your choice.
What other constituents would you need?

(3 points) 16) Write the Henderson-Hasselbalch equation for a buffer system.

(3 points) 17) Is this structural formula "correct" for physiological pH?

```
    H  O
    |  ||    //O
H — C — C — C
    |        \
    H         OH
```

If not, correct it and explain.

(3 points) 18) Without thinking adequately, your instructor asks you to prepare a laboratory solution containing: 100 mM NaCl, 10 mM $CaCl_2$, 10 mM NaH_2PO_4, and 10 mM Na_2HPO_4.
Knowingly, but gently, you warn him that. . . .

(72 points total)

SELF-TEST KEY UNITS 6, 7, & 8

(Criterion score 65 points: total of 72 points)

(4 points) 1) 3C–$3 \times 12 = 36$
3O–$3 \times 16 = 48$
4H– $4 \times 1 = 4$
M.W. = 88 daltons (Give partial credit for correct methods.)
(Review: Unit 6, Items 1-6)

(1 point) 2) 88 grams (1) *(Review: Unit 6, Item 7)*

(14 points) 3) Calculate: G.M.W. = 88 g
1 ℓ 1 M = 88 g/ℓ.
1 ℓ 0.2 M = 17.6 g/ℓ.
500 ml 0.2 M = 8.8 g/500 ml. (3 points)
Obtain a 500 ml volumetric flask. (1)
Weigh out 8.8 g of pyruvic acid, (1) and place it in the flask.
Add about 400 ml of distilled water. (1)
Swirl until dissolved. (1)
Add distilled water to the 500 ml mark. (1)
Mix thoroughly. (1) Place in a labeled (1) lab. bottle. (1)
Clean up! (3) *(Review: Unit 6, Items 9-14)*

(3 points) 4) 1% = 1 g/100 ml of solution
= 2 g/200 ml
3% = 6 g/200 ml (3 points) *(Review: Unit 6, Item 17)*

(3 points) 5) 1% = 1 ml/100 ml of solution
= 8 ml/800 ml
20% = 160 ml/800 ml of solution (2)
plus 640 ml distilled water (1) *(Review: Unit 6, Items 18-20)*

(3 points) 6) 1 ppm = 1 mg/liter of solution
2 ppm = 2 mg/ℓ
= 4 mg/2 ℓ (2)
Add about 2 ℓ of distilled water (2 mg of estrogen occupies negligible space). (1) *(Review: Unit 6, Item 21)*

(3 points) 7) 1 mg% = 1 mg/100 ml
950 mg% = 950 mg/100 ml
= 47.5 g/5 ℓ (3) *(Review: Unit 6, Item 22)*

(4 points) 8) $V \times M = V' \times M'$ (1)
$V \times 7 = 500 \times 0.3 = 150$
V = 150/7 = 21.4 ml of stock solution. (2)
Add 478.6 ml distilled water. (1) *(Review: Unit 6, Items 23-25)*

(1 point) 9) 0.8 mM (1) *(Review: Unit 6, Items 26-27)*

(4 points) 10) log 2 = 0.3
log 3 = 0.5
log 5 = 0.7
log 8 = 0.9 *(Review: Unit 7, Item 19)*

(3 points) 11) $pH = -\log(H^+)$ $\left(\text{or } \log \frac{1}{(H^+)}\right)$. (3) *(Review: Unit 7, Items 24 and 25)*

(5 points) 12) $pH = -\log(H^+) = -\log 0.008$

$= -\log(8 \times 10^{-3})$

$= -(\log 8 + \log 10^{-3})$

$= -(0.9 - 3) = -(-2.1)$

$= 2.1$ (5: give partial credit for correct method)

(Review: Unit 7, Items 26–28)

(5 points) 13) $pH + pOH = 14$

$pOH = 14 - 12 = 2$ (1)

$2 = -\log(OH^-)$ (1)

$10^{-2} = (OH^-)$

$(OH^-) = 0.01$ M (1)

G.M.W. = 40 g (1)

Use 0.4 g (1) *(Review: Unit 7, Items 29–33)*

(5 points) 14) $pH = pK + \log \frac{(A^-)}{(HA)}$ (2)

$5.25 = 4.75 + \log \frac{(A^-)}{(HA)}$

$0.5 = \log \frac{(A^-)}{(HA)}$

$3 = \frac{(A^-)}{(HA)} = \frac{\text{dissoc.}}{\text{undissoc.}} = \frac{3}{1}$ (3) *(Review: Unit 8, Items 1–4)*

(5 points) 15) Succinic (2)

pK is closest to the desired pH. (2)

A salt of the succinic acid (for example, sodium succinate). (1)

(Review: Unit 8, Items 13–30)

(3 points) 16) $pH = pK + \log \frac{(\text{salt})}{(\text{acid})}$ (3 points) *(Review: Unit 8, Item 26)*

(3 points) 17) No. (1) $H-\overset{H}{\underset{H}{C}}-\overset{O}{\overset{\|}{C}}-C(=O)O^{\ominus}$ (1) *(Review: Unit 8, Item 32)*

Weak organic acids (carboxyl groups) are dissociated at physiological pH. (1)

(3 points) 18) Calcium phosphate will precipitate out. (1)

In the presence of Ca^{++}, the total phosphate should not exceed 3mM at pH = 7.2. (1)

Hence either the phosphate should be reduced or the calcium should be eliminated. (1) *(Review: Unit 8, Items 33 and 34)*

Take at least a 5-minute break before continuing with the next unit.

UNIT NINE 9

CARBOHYDRATES: MONOSACCHARIDES AND RELATED MOLECULES

CONCEPTS

Carbohydrates are one of the four major classes of biochemical molecules. They are the chief constituents of the metabolic reaction sequences of photosynthesis, glycolysis, and Krebs cycle and also function as energy storage molecules. In addition, carbohydrates and their derivatives form the architectural molecules of plant and bacterial cell walls, insect cuticles, and the "intercellular cement" of all animal cells. You will be asked to learn the names and structural formulae for several of the biologically important monosaccharides and related molecules. When you study the complex reaction sequences of photosynthesis and respiration, these processes should be easier to follow and to understand if you remember the material from this unit and the next one.

A carbohydrate molecule = carbon + water = approximately $C_n(H_2O)_n$

All sugars are carbohydrates; for example, glucose, galactose, fructose, sucrose, lactose, etc.

A monosaccharide, or simple sugar, is a carbohydrate that cannot be broken down into smaller carbohydrates by treatment with acids. The most common monosaccharides are hexoses and pentoses.

Common Hexose Monosaccharides

	Glucose	Galactose	Fructose
Empirical formulae	$C_6H_{12}O_6$	$C_6H_{12}O_6$	$C_6H_{12}O_6$
Structural formulae	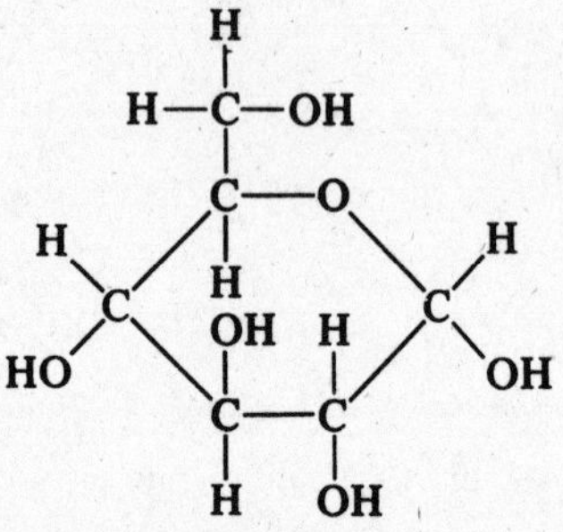	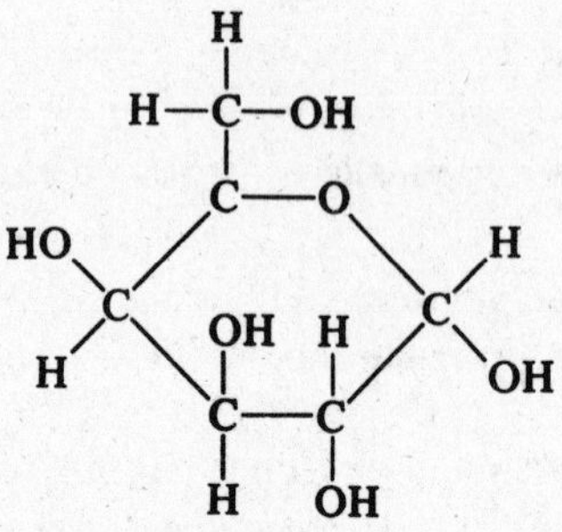	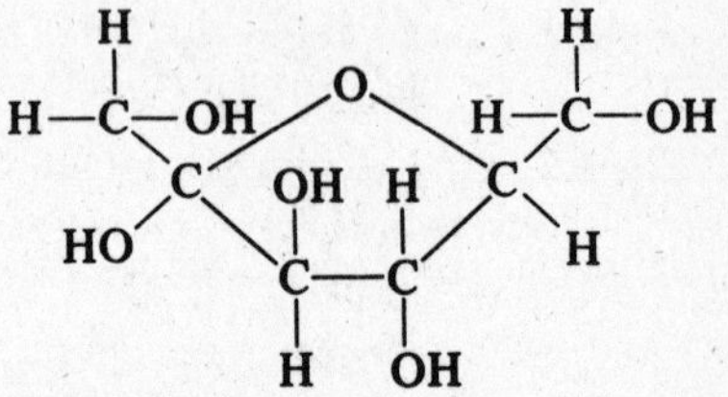
Outline structural formulae			

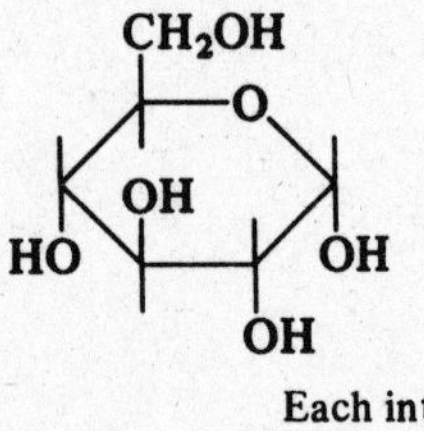

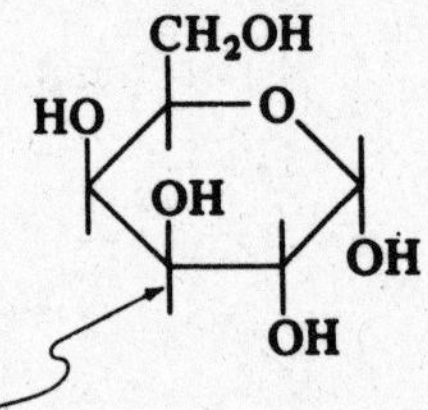

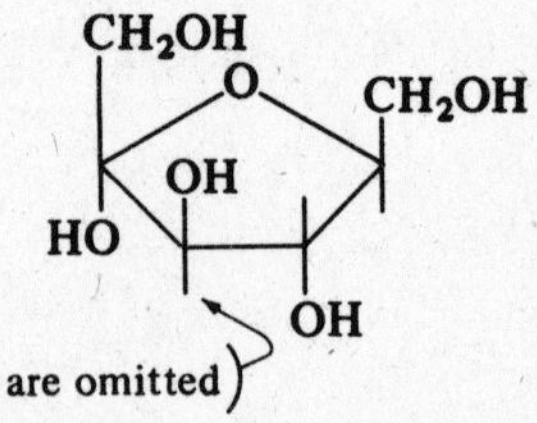

Each intersection is a C atom

H atoms are omitted

In hexose structural formulae the carbons are numbered thus:

```
 ⑥C
  |
 ⑤C—O
  /    \
④ C      C ①
  \    /
   C—C
   ③   ②
```

Naming Sugars

Sugar	Carbon Length	Latin or Greek Number
Triose	3	Treis
Tetrose	4	Tetra
Pentose	5	Penta
Hexose	6	Hex
Heptose	7	Hepta
Octose	8	Octa

QUESTIONS AND PROBLEMS / SAMPLE RESPONSES

1) Carbohydrates: Carbo + hydrate (carbon + water)
Empirical formula: $C_n(H_2O)_n$
A 3-carbon carbohydrate: $C_3H_6O_3$
A 4-carbon carbohydrate: $C_4H_8O_4$
What would be the empirical formula for a 5-carbon carbohydrate?

—$C_5H_{10}O_5$

2) Write the empirical formula for a 6-carbon carbohydrate.

—$C_6H_{12}O_6$ *Note:* there are some exceptions to this general formula $C_n(H_2O)_n$

3) Sugars are carbohydrates.
Sugars: Glucose
Ribose
Fructose
What would you guess to be the name for malt sugar?

—Maltose

④ What word ending is characterististic for sugars?

—-ose

⑤ Complete the table:

Sugar	Number of Carbon Atoms	Greek or Latin Number	
Triose	3	Treis	
Tetrose	4	Tetra	
	5	Penta	—Pentose
	6	Hex	Hexose
	7	Hepta	Heptose
	8	Octa	Octose

6) How many carbons does a hexose have?

—6

CONCEPTS: THREE CARBON COMPOUNDS

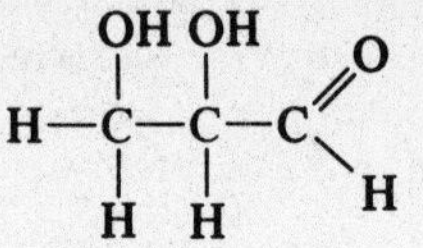

Glyceraldehyde

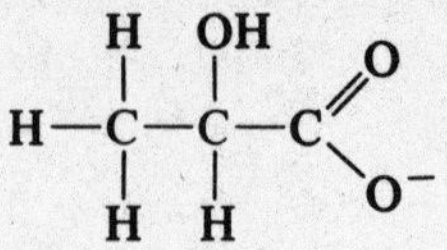

Lactate
(Lactic acid)

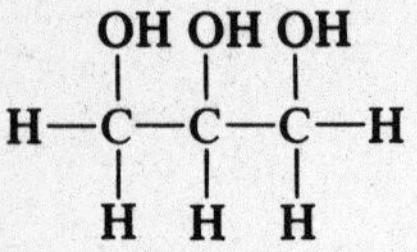

Glycerol

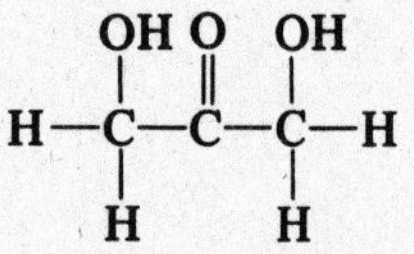

Dihydroxyacetone

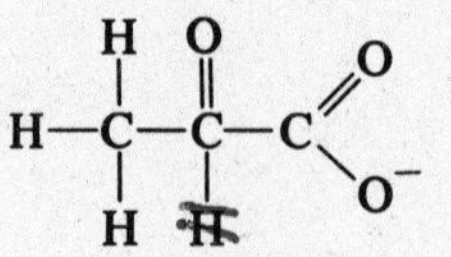

Pyruvate
(pyruvic acid)

Glycerol is a constituent of all fats and phospholipids.

Glyceraldehyde, dihydroxyacetone, and pyruvic acid are intermediate molecules in the metabolic breakdown of glucose in all cells.

Lactic acid is a temporary product of glucose metabolism in muscle cells during vigorous exercise.

QUESTIONS AND PROBLEMS

SAMPLE RESPONSES

7) All the following are carbohydrates:

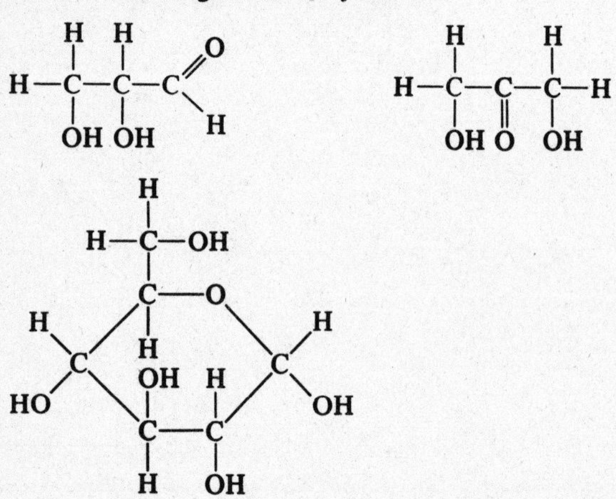

What chemical group (studied in Unit 2) is found in *all* carbohydrates? What other group is found in the first (at the left)? What other group is found in the second?

– —OH (hydroxyl)

(aldehyde)

C—C—C (keto)

8) Another definition of simple carbohydrates is that they are polyhydroxyaldehydes or polyhydroxyketones. Label each of these correctly and give its specific name from Concepts.

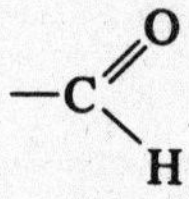

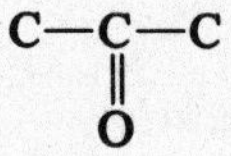

A.

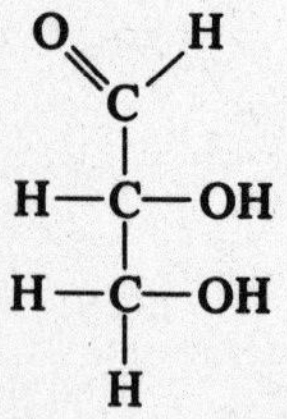

B.

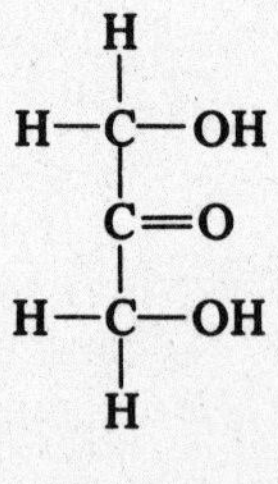

–A. Polyhydroxyaldehyde, glyceraldehyde
B. Polyhydroxyketone, dihydroxyacetone

9) Pyruvic acid and lactic acid are related to the trioses. *Pyruvic acid* contains a keto group; *lactic acid* does not. Name each of these molecules.

```
A.      H                B.    O⊖    O
        |                        \  //
    H—C—OH                        C
        |                         |
       C=O                       C=O
        |                         |
    H—C—OH                    H—C—H
        |                         |
        H                         H

C.   O     O⊖            D.   H     O
      \\  /                     \  //
        C                         C
        |                         |
    H—C—OH                    H—C—OH
        |                         |
    H—C—H                     H—C—OH
        |                         |
        H                         H
```

–A. Dihydroxyacetone
B. Pyruvate (puruvic acid)
C. Lactate (lactic acid)
D. Glyceraldehyde

10) The formula for glyceraldehyde is $CH_2OHCHOHCHO$. Write the structural formula.

```
   OH OH
    |  |     O
–H—C—C—C//
    |  |    \
    H  H     H
```

11) The formula for dihydroxyacetone is $CH_2OHCOCH_2OH$. Write the structural formula.

```
   OH    OH
    |     |
–H—C—C—C—H
    |  ||  |
    H  O  H
```

12) From memory, write the structural formulae for gluteraldehyde and dihydroxyacetone.

```
    H  H                      H  O  H
    |  |     O                |  ||  |
–H—C—C—C//                H—C—C—C—H
    |  |    \                 |     |
   OH OH     H                OH    OH
```

13) The formula for lactic acid is $CH_3CHOHCO_2H$ or $CH_3CHOHCOOH$. Write the structural formula.

```
    H  OH
    |  |     O
–H—C—C—C//
    |  |    \
    H  H     OH
```

14) Write the structural formula for lactate without consulting the preceding items.

```
    H  H
    |  |     O
–H—C—C—C//
    |  |    \
    H  OH    O⁻
```

15) The formula for pyruvic acid is CH_3COCO_2H or $CH_3COCOOH$. Write the structural formula for pyruvic acid.

```
    H  O
    |  ||    O
–H—C—C—C//
    |       \
    H        OH
```

(Pyruvic acid is a three-carbon intermediate in the metabolic breakdown of glucose in almost all cells.)

16) Write the structural formula for pyruvate.

```
    H  O
    |  ||    O
–H—C—C—C//
    |       \
    H        O⁻
```

17) Is glucose a hexose or a pentose? Does glucose conform to the general formula for carbohydrates ($C_n(H_2O)_n$)?

–Hexose
–Yes ($C_6H_{12}O_6$)
(Glucose is one of the fundamental foods–fuels–for the cells of higher organisms; both plants and animals.)

18) The carbons are numbered thus:

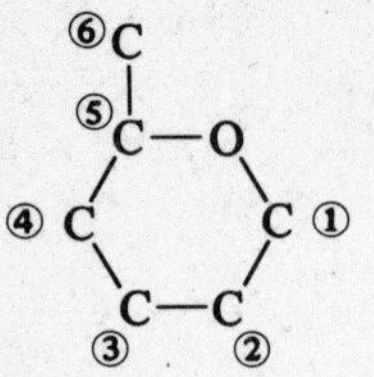

Without consulting the preceding diagram, copy this "skeleton" and try to complete the structural formula for glucose.

– (Note: This –OH can be in any position.)

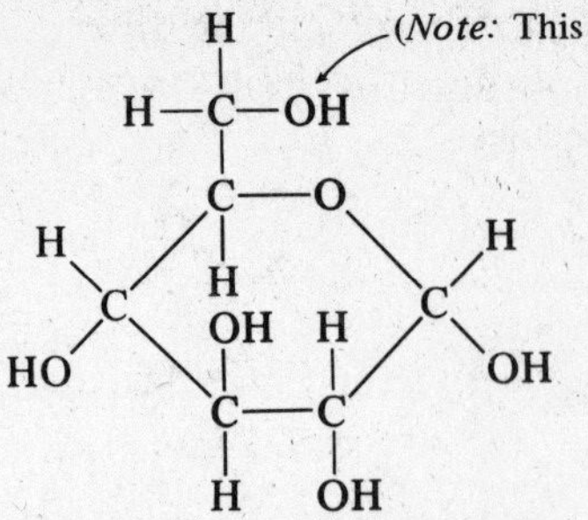

Correct any errors you made. Note that the H and OH are "inverted" on carbon #3. (This is an oversimplification of the rather complex phenomenon of stereoisomerism.)

19) Write the structural formula for glucose.

–

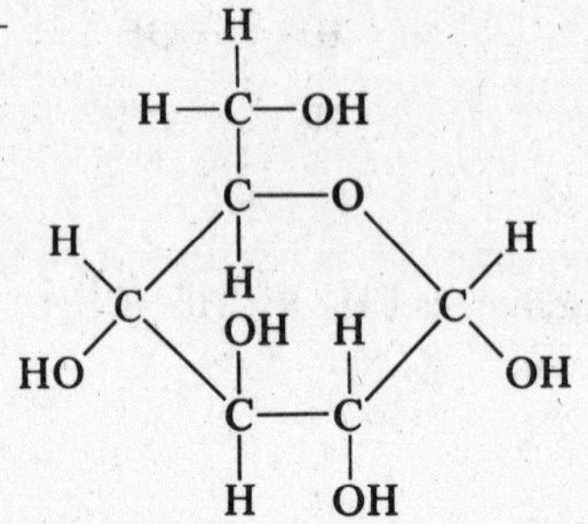

20) Which carbon (by number) is not in the glucose "ring"? The glucose ring is usually written in "shorthand" form:

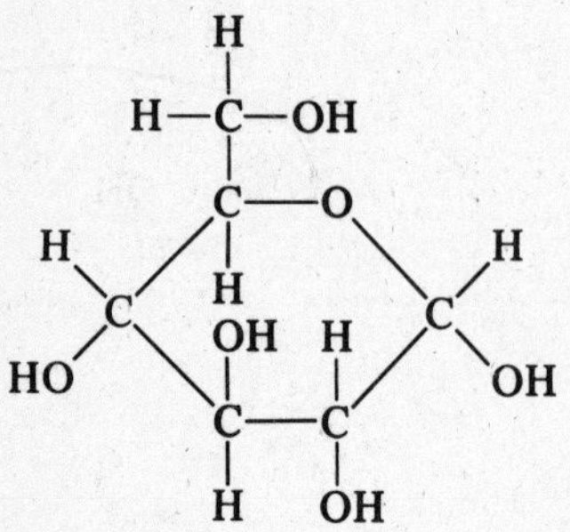

or

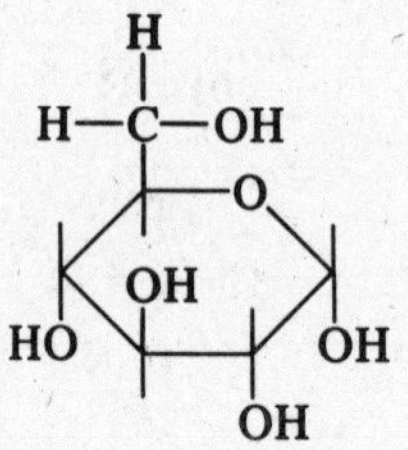

or

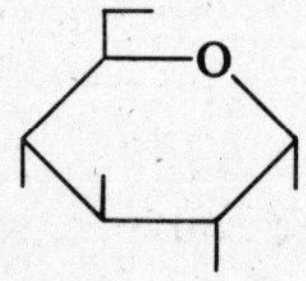

In the lower diagram, what does the (┌) represent?
In the lower diagram, what does the (|) represent?

– #6

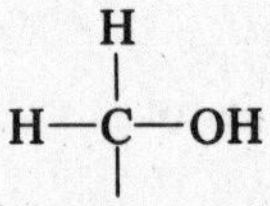

– H–C–OH or CH_2OH

– –OH

21) Complete this skeleton by adding OH groups and a CH_2OH group to make a "shorthand" glucose formula.

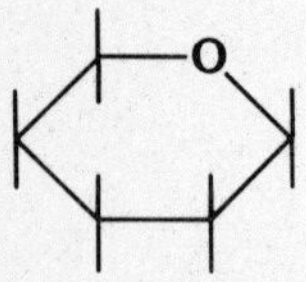

–

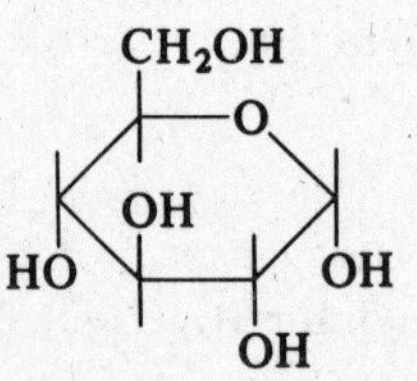

22) Write a "shorthand" formula for glucose.

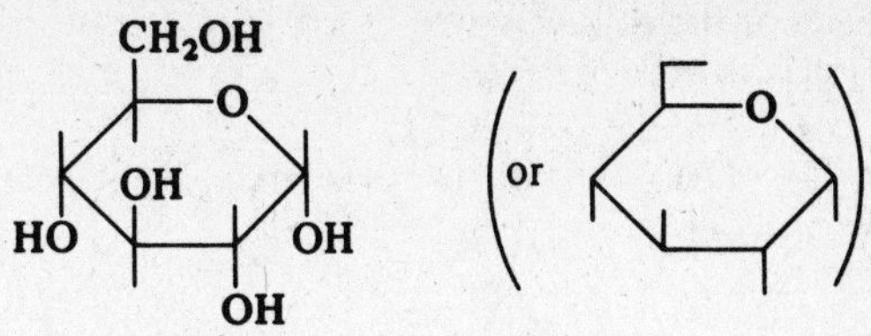

23) What is the difference between galactose and glucose?

–Inverted OH on carbon #4

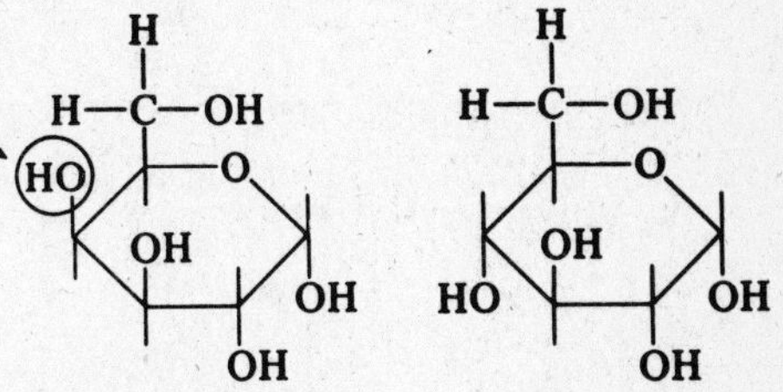

Galactose

Glucose

(Galactose and glucose can combine to form lactose, the characteristic sugar of milk.)

24) Write the "shorthand" structural formula for fructose.

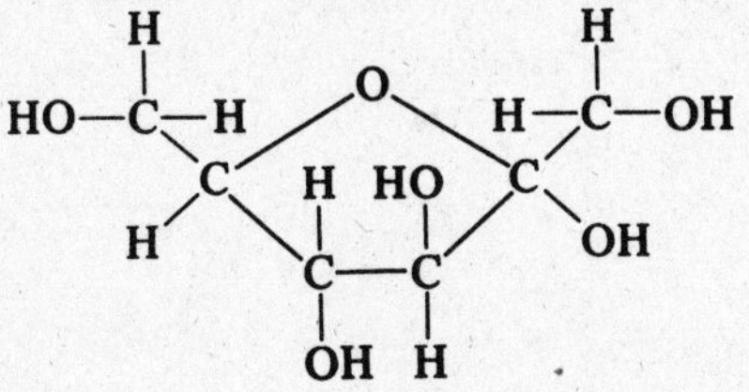

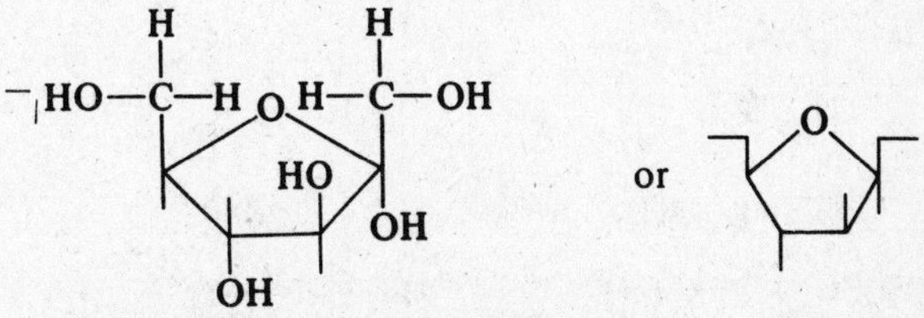

(Fructose is the sugar that is characteristic of fruit; hence, it is commonly called "fruit sugar." It is also an intermediate in respiration and photosynthesis.)

25) Which carbon has an "inverted" OH group?

–Carbon #3

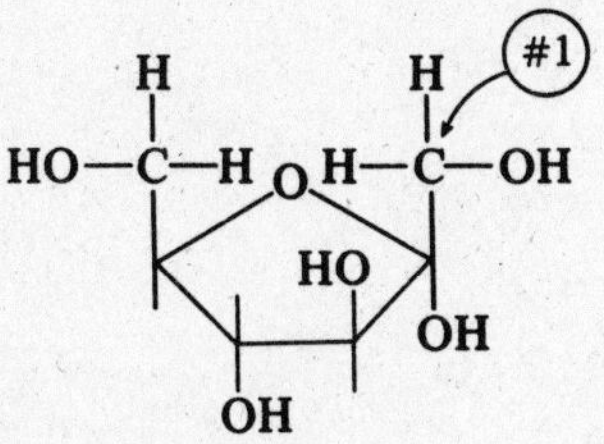

26) Write the "shorthand" structural formula for fructose.

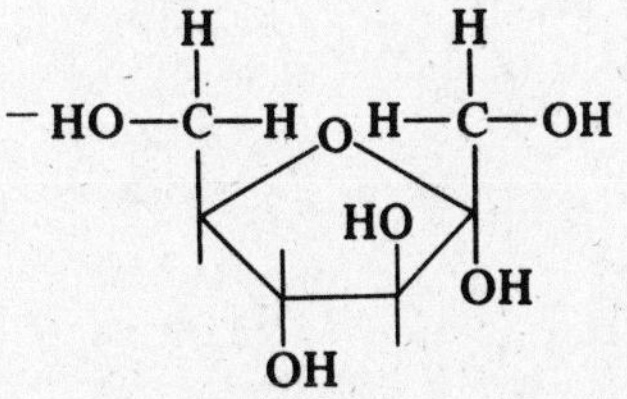

(27) Name each of the following:

A.

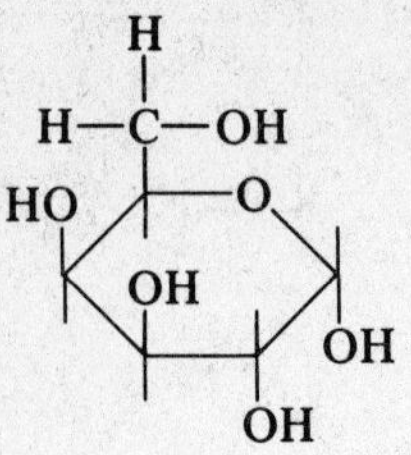

B.

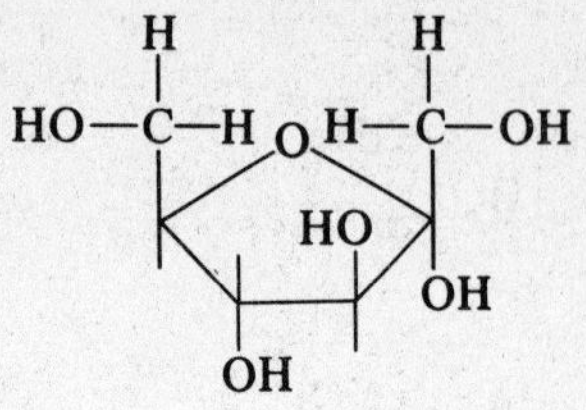

C.

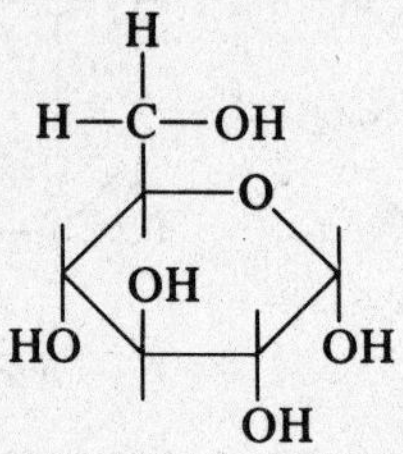

–A. Galactose
B. Fructose
C. Glucose

(28) Indentify each of the following:

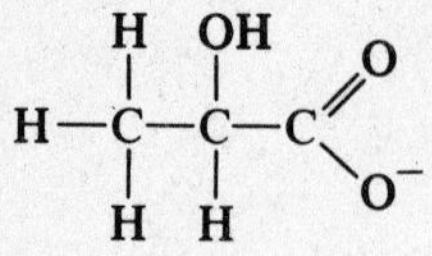

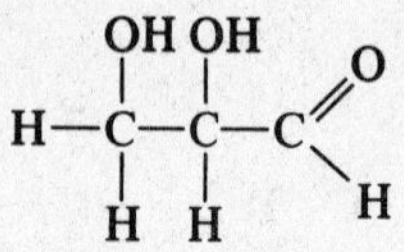

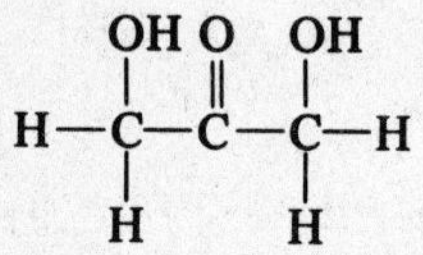

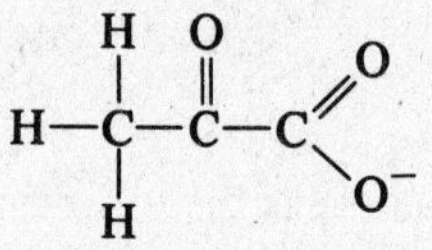

–Lactate
Dihydroxyacetone
Glyceraldehyde
Pyruvate

(29) Without consulting preceding items, write the structural formula for glucose.

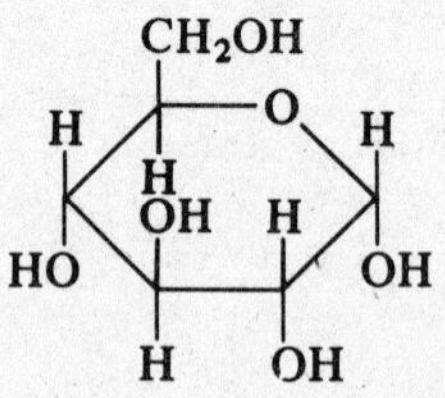

Take at least a 5-minute break before going on to the next unit.

UNIT TEN 10

PENTOSES, DISACCHARIDES, AND POLYSACCHARIDES

Ribose and deoxyribose are constituents of the nucleic acids RNA and DNA, respectively.

Ribose

Deoxyribose

Alpha glucose and beta glucose are structurally very similar, but they are chemically quite different.

α-glucose

β-glucose

Each disaccharide is composed of two monosaccharides.

Sucrose

α-glucose and fructose

Cellobiose

β-glucose and β-glucose

Maltose

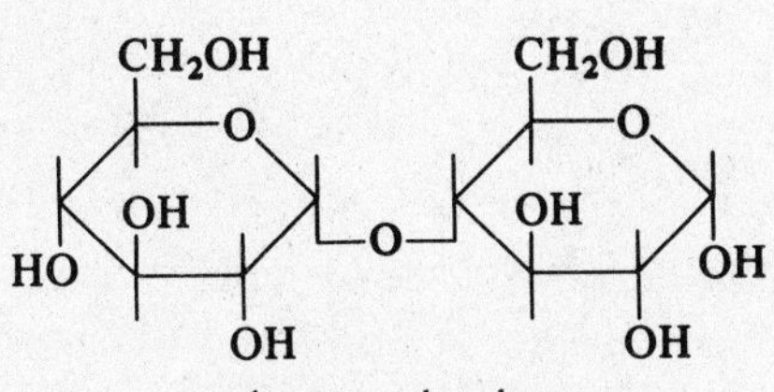

α-glucose and α-glucose

Lactose

β-galactose and α-glucose

QUESTIONS AND PROBLEMS

1) How many carbons do ribose and deoxyribose each have?
What general name would apply to both?
What is the difference between the structural formulae of ribose and deoxyribose?

Which carbon has an inverted OH group in the ribose structural formula?

SAMPLE RESPONSES

–5
–Pentose

–Deoxyribose lacks an OH group on carbon #2. (OH is replaced by H).

–Carbon #1.
Pronunciation: RYE′-BOSE (BOSE rhymes with glucose). Say it *out loud* three times.

2) Write the "shorthand" structural formulae for ribose and deoxyribose.

–

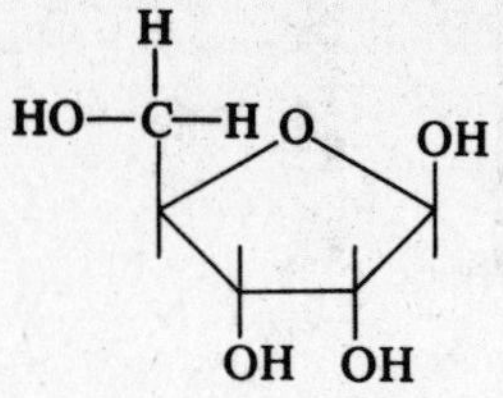

Ribose

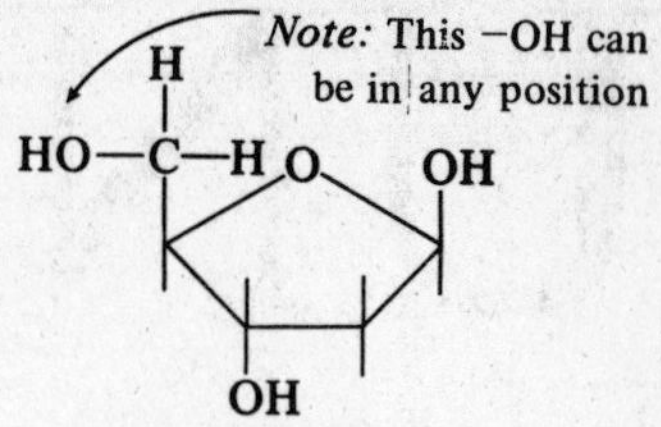

Note: This –OH can be in any position

Deoxyribose

3) Write an outline structural formula for each of the following:
Glucose
Ribose
Deoxyribose

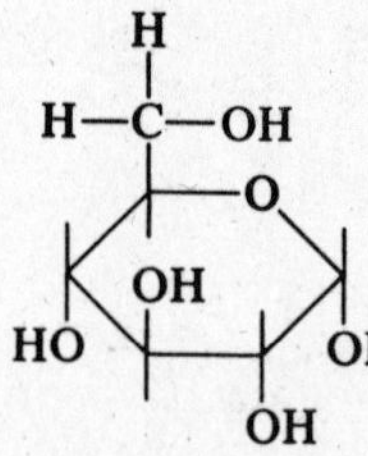

Glucose

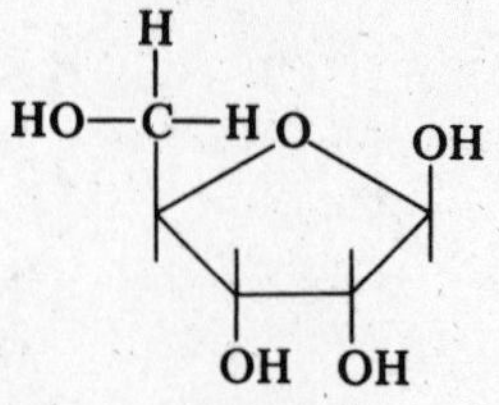

Ribose

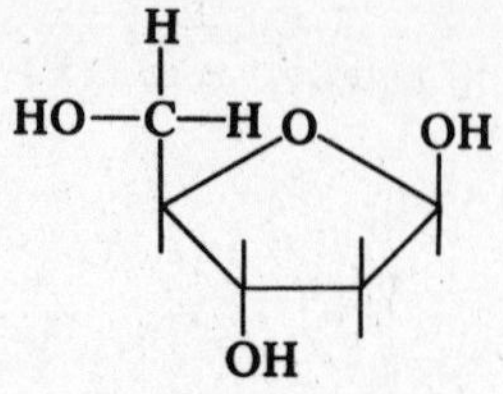

Deoxyribose

4)

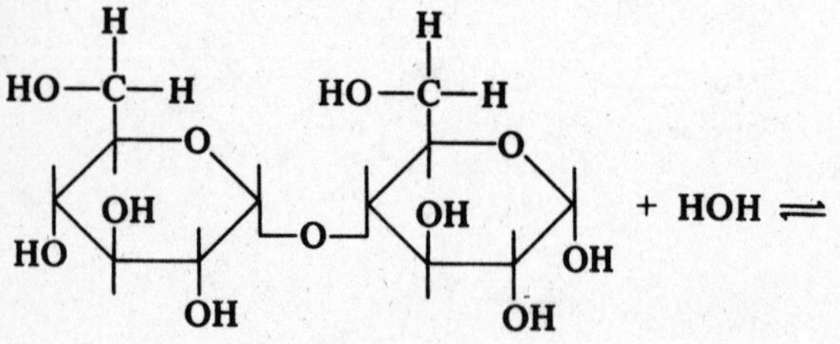

Maltose + HOH ⇌ glucose + glucose

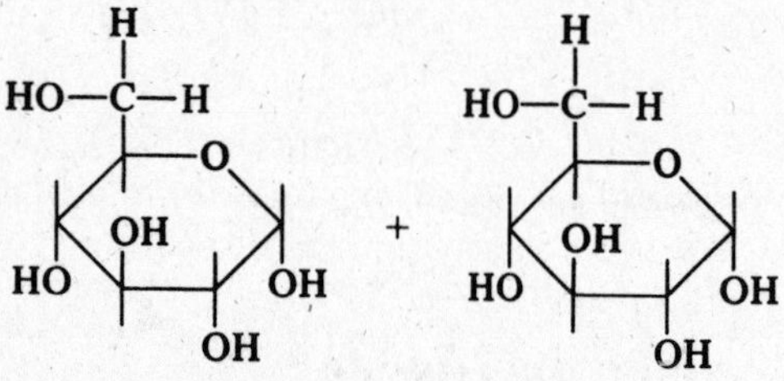

This reaction, from left to right, is an example of a *hydrolysis* reaction. What molecule is always added in a hydrolysis reaction?

–H_2O (HOH or water)

What two molecules are produced by the hydrolysis of maltose?

–2 molecules of glucose

5) Similarly:
2 Glucose ⇌ _______ + _______
Fill in the blanks.

–Maltose + water (H_2O)

6) Glucose is a *mono*saccharide. Maltose is composed of two monosaccharides (2 glucose). What general term would you apply to maltose?

–*Di*saccharide

7)

Sucrose

When sucrose is hydrolyzed, what two molecules are produced?

–Glucose + fructose

8) What is the difference between the structural formulae for alpha glucose and beta glucose?

–β glucose has the H and OH attached to carbon No. 1 inverted. (or similar response) (This is also an oversimplification of the phenomenon of stereoisomerism. Although they appear to be almost identical, α and β glucose behave quite differently biochemically.)

9)

α (alpha) linkage

Maltose

β (beta) linkage

Cellobiose

What two molecules will cellobiose produce upon hydrolysis?

–2β glucose

What is the difference between the α and β linkage?

–α linkage–2α glucose; β linkage–2β glucose. Also, in the β linkage the second glucose is inverted (upside down). (or similar response) (Cellobiose is the disaccharide building block of cellulose.)
Pronunciation: CELL-O-BI′-OSE

CONCEPTS

A polysaccharide is composed of three or more monosaccharides.

Starches are polysaccharides composed of alpha glucose monosaccharide units. They function as energy storage molecules.

Amylose

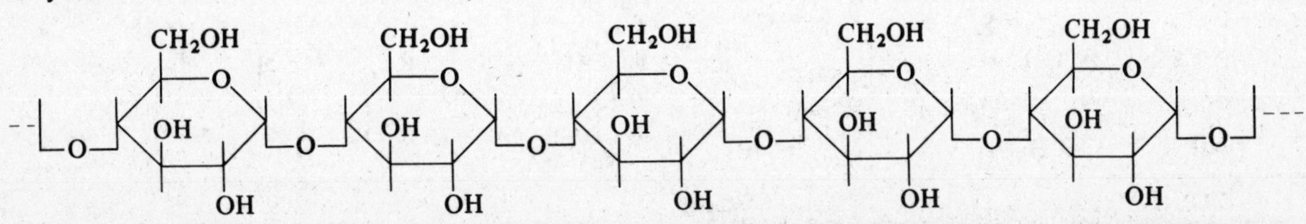

Alpha glucose units can also form 1,6 linkages that are characteristic of amylopectin in plants and glycogen in animals.

1-6 linkages: amylopectin and glycogen

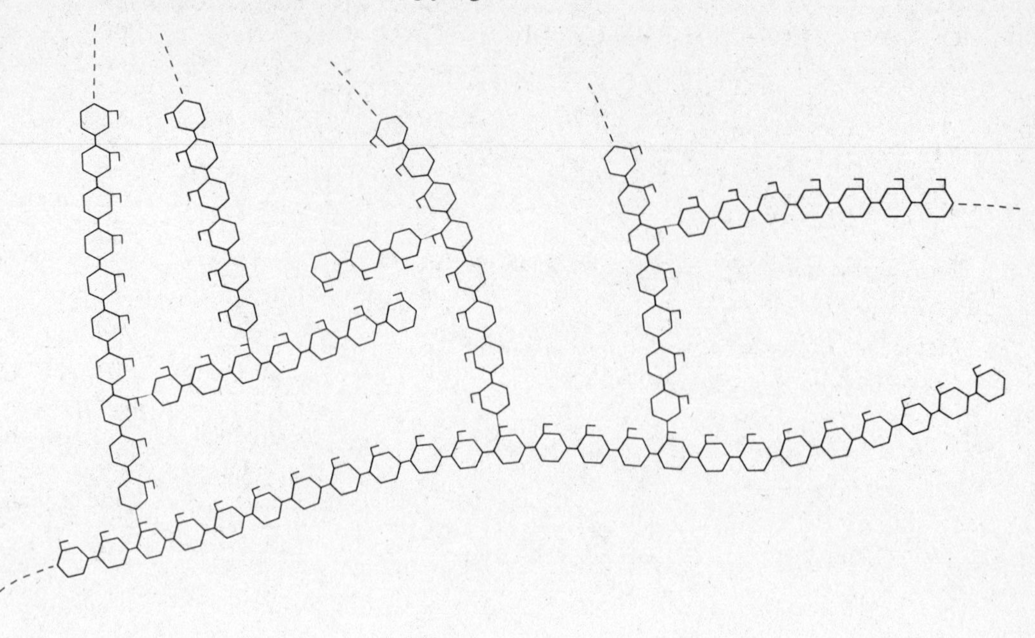

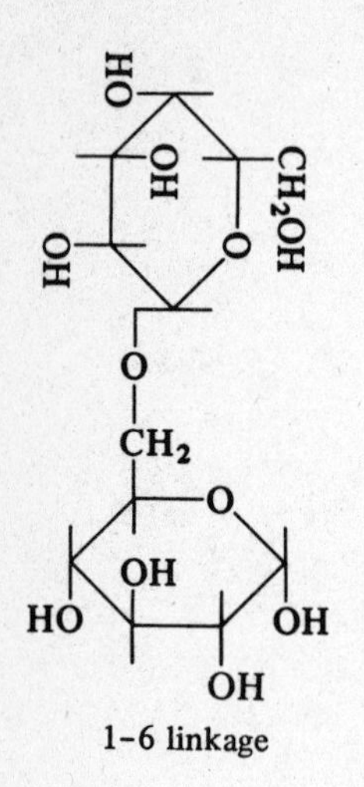

1-6 linkage

Cellulose is the major constituent of plant cell walls, which also contain other, more complex polysaccharides.

Cellulose

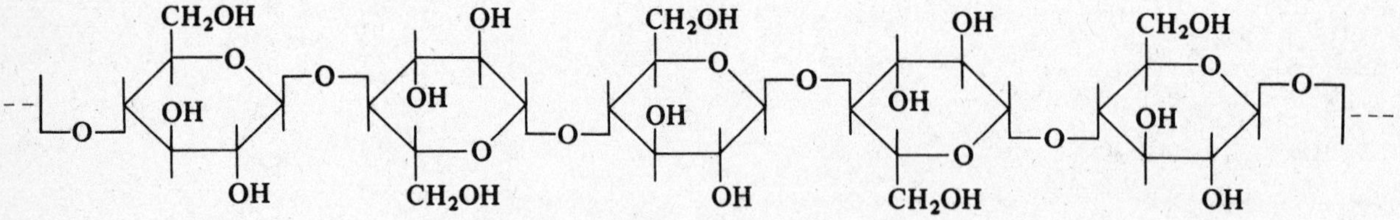

CH_2OH OH CH_2OH OH CH_2OH

OH OH OH OH OH

OH CH_2OH OH CH_2OH OH

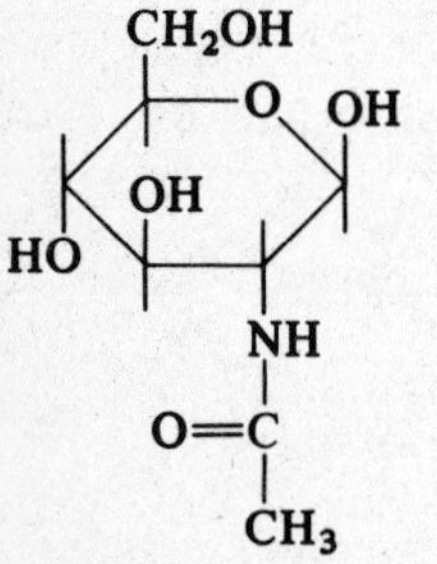

CH_2OH OH HO NH $O{=}C$ CH_3

β N-acetyl glucosamine

Chitin is the major constituent of insect and other arthropod exoskeletons. It is composed of beta N-acetyl-glucosamine units.

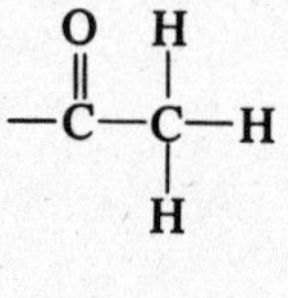

$$-\overset{\overset{\large O}{\|}}{C}-\underset{\underset{\large H}{|}}{\overset{\overset{\large H}{|}}{C}}-H = -COCH_3 = \text{an acetyl group}$$

an N-acetyl group is bonded to a nitrogen

Chitin

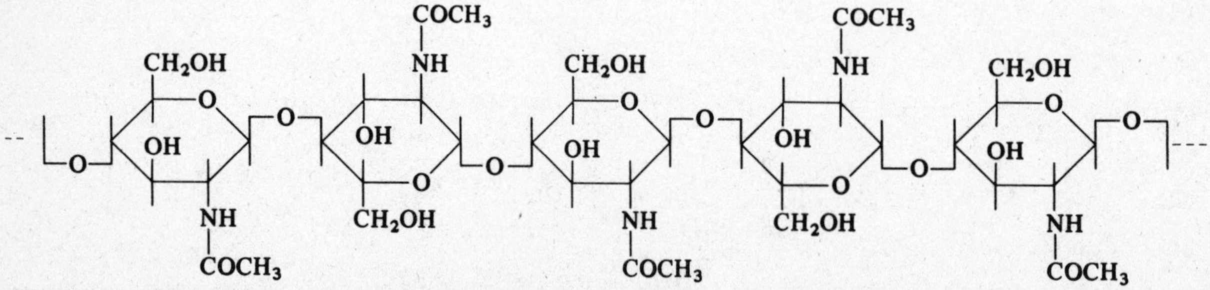

$COCH_3$ $COCH_3$

CH_2OH NH CH_2OH NH CH_2OH

OH OH OH OH OH

NH CH_2OH NH CH_2OH NH

$COCH_3$ $COCH_3$ $COCH_3$

QUESTIONS AND PROBLEMS / **SAMPLE RESPONSES**

10)

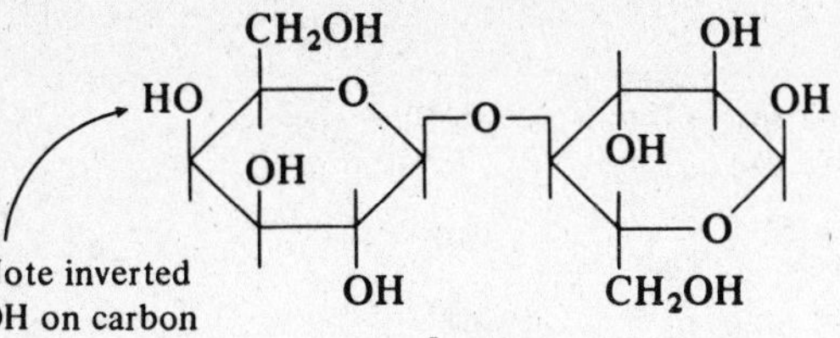

Lactose

Note inverted OH on carbon No. 4

What two monosaccharides will be produced by hydrolysis of lactose? What type of linkage is involved?

–Glucose + galactose (actually α glucose and β galactose)
–β linkage

(11) What two monosaccharides are produced upon hydrolysis of each of the following disaccharides?
Sucrose
Maltose
Lactose
Cellobiose
Which two contain β linkages?

–Sucrose: Glucose + fructose
Maltose: 2 glucose
Lactose: glucose + galactose
Cellobiose: 2 glucose
–β linkages: lactose and cellobiose

12) Monosaccharide = a simple sugar
Disaccharide = 2 simple sugars
Polysaccharide =

–Many simple sugars

13) *Starch* functions in food storage in plants and animals. What general term may be applied to starch? Of what monosaccharide units is starch composed? What type of linkage is involved?

–Polysaccharide
–Glucose
–α

14) In the diagram of a starch molecule, which carbons of glucose are linked through O?

–The #1 and #4 carbons

15) This is a 1,4 α linkage and is characteristic of *amylose*–a straight chain starch. Glucose units can also be linked by a 1,6 α linkage. Add another glucose to the molecule below by a 1,6 α linkage.

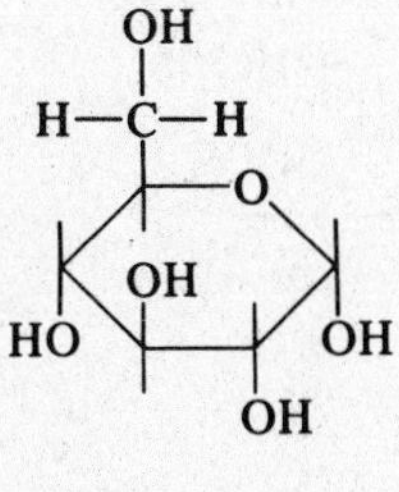

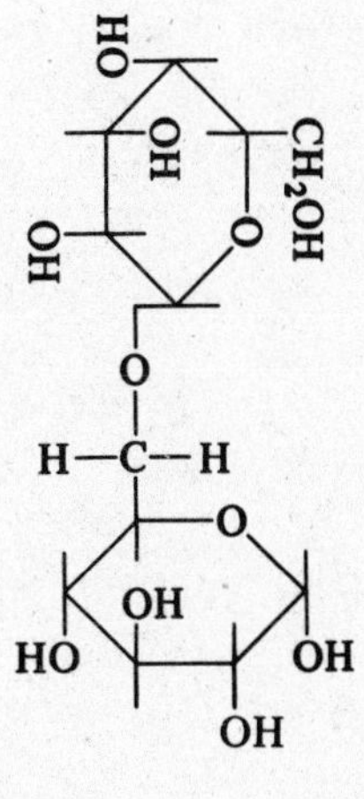

– or

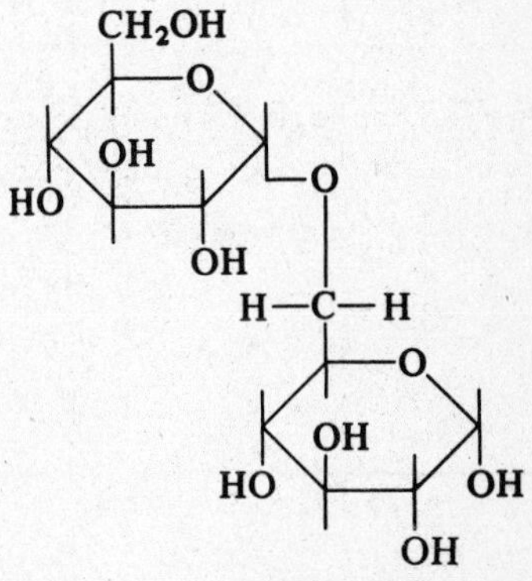

(or similar response)

16) The linkage you have just written is characteristic of *amylopectin* in plants and *glycogen* in animals. What linkage is characteristic of:

Amylose____________

Amylopectin and glycogen____________

–Amylose–1,4 α linkage
–Amylopectin and glycogen–1,6 α linkage (in addition to 1,4 α linkages)

17) Label the following either amylose or amylopectin and glycogen.

–A. Amylopectin and glycogen
B. Amylose

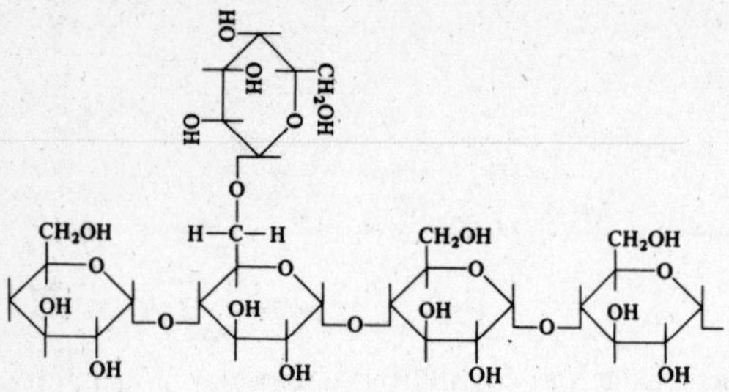

A. (30 or less glucose units)

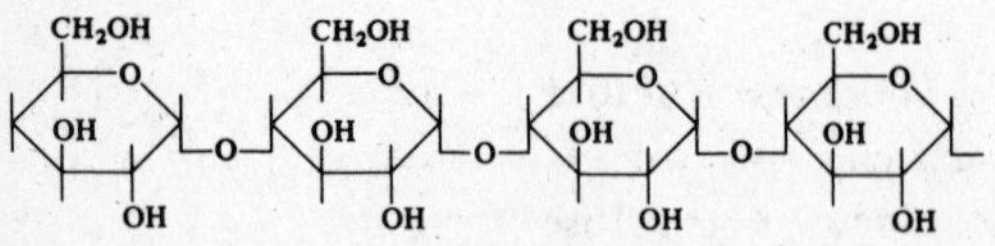

B. (about 1000 glucose units)

18) Cellulose–constituent of plant cell walls.

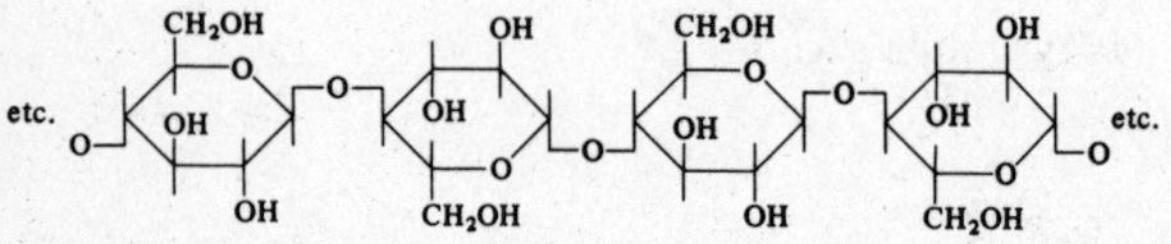

Of what monosaccharide units is cellulose composed?

–β glucose

What is the difference between cellulose and starch?

–Cellulose has β linkages (*alternative glucose units are inverted*); starch has α linkages

Of what disaccharide units is cellulose composed?

–Cellobiose

(19) *β glucosamine* is a derivative of β glucose in which the OH group of carbon #2 is replaced by an amino group. Write the "shorthand" structural formula for glucosamine.

Then replace one of the amino hydrogens with an acetyl group to make N-acetyl glucosamine.

–

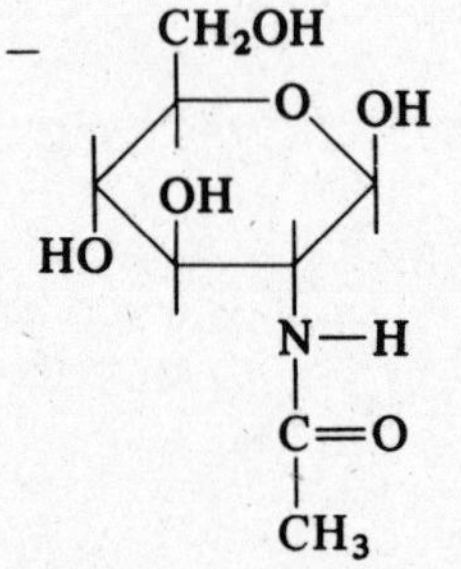

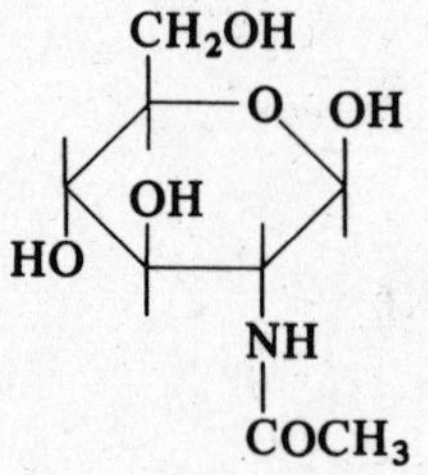

20) Chitin–a constituent of *arthropod cuticle.*

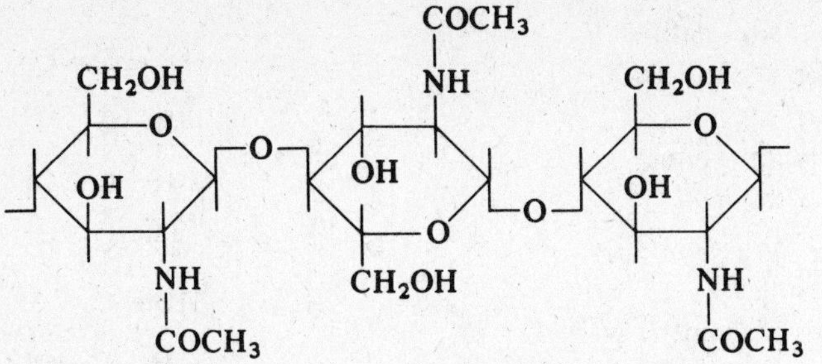

Of what units is chitin composed? What type of linkage is involved?

–β N-acetylglucosamine
–β linkages

(21) Complete the table:

Polysaccharide	Units	Linkages	Biological Function or Location
Amylose			
Glycogen			
Chitin			
Cellulose			
Amylopectin			

–Polysaccharide	Units	Linkages	Biological Function or Location
Amylose	Glucose (ca. 1000)	1,4 α	Plant food storage
Glycogen	Glucose (30 or less)	1,4 α and 1,6 α	Animal food storage
Chitin	N-acetylglucosamine	β (1,4)	Arthropod cuticle
Cellulose	Glucose (β)	β (1,4)	Plant cell wall
Amylopectin	Glucose	1,4 α and 1,6 α	Plant food storage

(If you did not fill all the blanks, look up the appropriate information from preceding items and complete the table.)

CONCEPTS: VITAMIN C, MUCOPOLYSACCHARIDES, GLYCOPROTEINS, BACTERIAL CELL WALLS

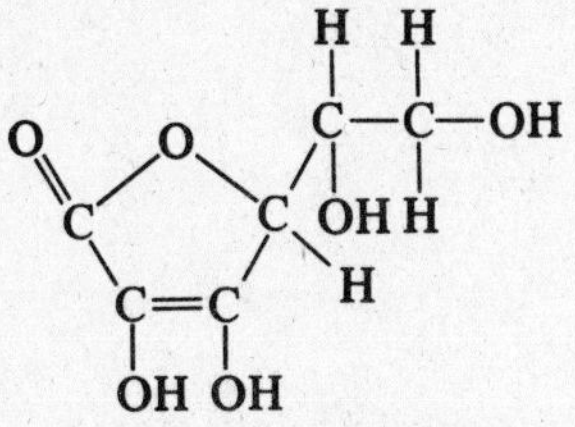

Ascorbic acid

Ascorbic acid is vitamin C, which functions in connective tissue formation, wound healing, and disease resistance.

Mucopolysaccharides are gelatinous, slimy, or sticky molecules composed of units that are related to hexoses. They function as intercellular cement and in connective tissues.

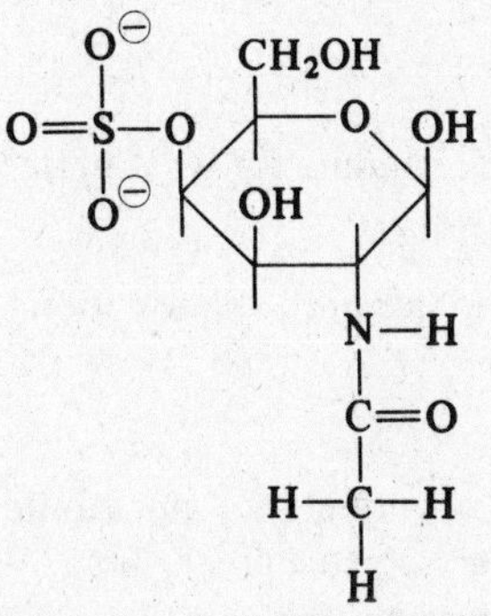

N-acetyl galactosamine sulfate

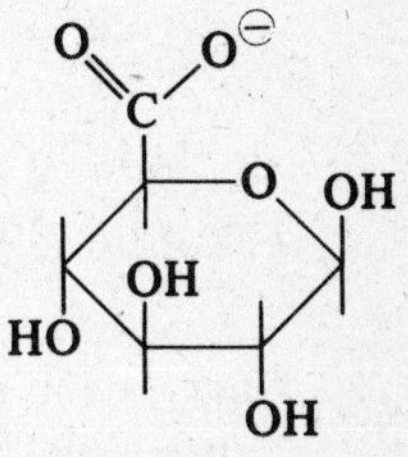

glucuronic acid

Mucopolysaccharide	Units	Functions
Chondroitin sulfate	Glucuronic acid, N-acetylgalactosamine sulfate	Matrix of cartilage and bone; outer coating on cell membranes.
Hyaluronic acid	Glucuronic acid, N-acetylglucosamine	Intercellular cement; synovial fluid of joints; vitreous humor of the eye.
Heparin	Glucosamine sulfate, glucuronic acid sulfate	Prevents blood clotting; lines blood vessels.

Glycoproteins are molecules in which carbohydrates are covalently bonded to proteins. They are of wide occurrence and functions: for example, cell membrane proteins, many blood proteins, follicle stimulating hormone, thyroid stimulating hormone, some enzymes, egg albumin, salivary gland secretions, stomach mucus, collagen, the lens capsule, blood type antigens (A, B, O), and "antifreeze" glycoproteins of arctic fishes. The short polysaccharide chains of glycoproteins commonly contain N-acetylneuraminic acid (sialic acid) along with other monosaccharides and related molecular subunits.

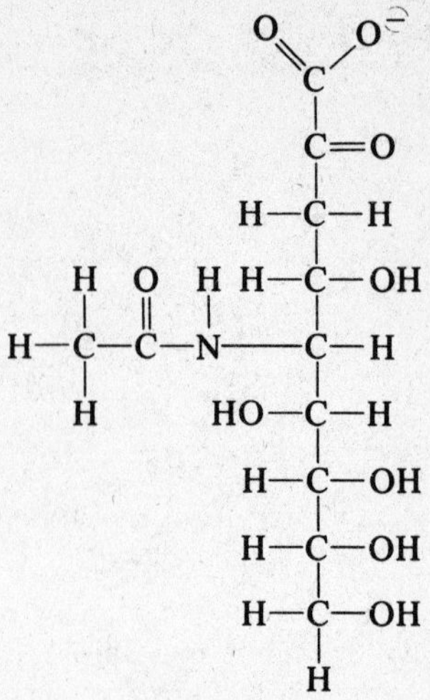

Bacterial cell walls are made of a complex molecular lattice that includes N-acetylglucosamine, N-acetylmuramic acid, and several amino acids. The entire wall, or capsule, is a single huge interconnected molecule, called a peptidoglycan. Gram-negative bacteria (those that do not stain with Gram stain) also contain many lipid molecules in their cell walls.

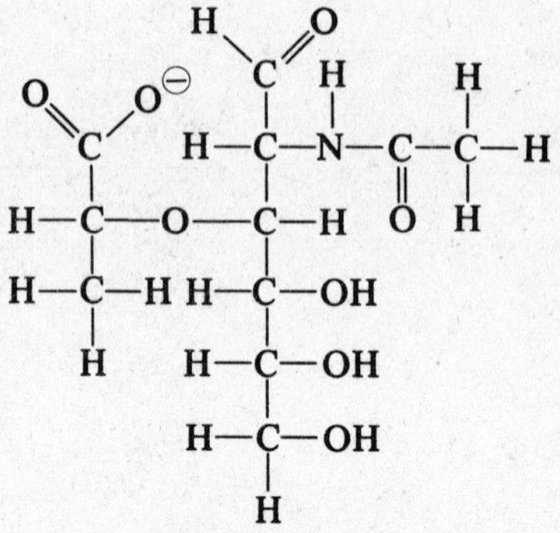

N-acetylneuraminic acid

N-acetylmuramic acid

QUESTIONS AND PROBLEMS

SAMPLE RESPONSES

22) This is the structural formula for what molecule?

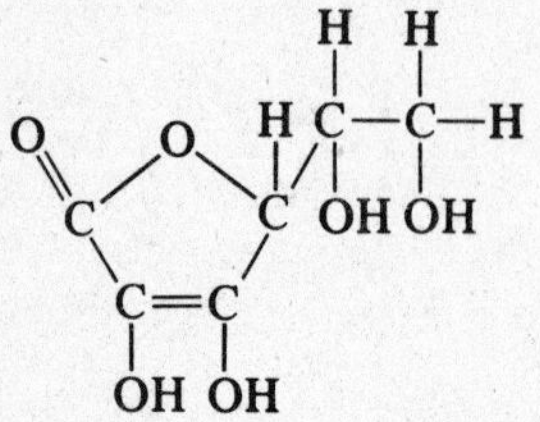

–Ascorbic acid (vitamin C)
Pronunciation: ASS-CORE′-BICK

23) What are the three major probable functions of ascorbic acid?

–Connective tissue formation, wound healing, and disease resistance.

24) Inspect the structural formula for glucuronic acid.
Is the molecule diagrammed alpha or beta?
What group is attached to carbon #5?
From what monosaccharide is glucuronic acid "derived"?

–Beta (β)
–Carboxyl (organic acid)
–β glucose
Pronunciation: GLUE-QUE-RON′-IC

25) Inspect the structural formula for N-acetyl galactosamine sulfate.
Is the molecule alpha or beta?
$-COCH_3$ is an acetyl group. To what carbon is the N-acetyl attached?
To what carbon is the sulfate group attached?

–beta
–#2
–#4
Pronunciation: ASSET-TEAL′

(26) Of what subunits is the polysaccharide chondroitin sulfate constructed?

–Glucuronic acid and N-acetyl galactosamine sulfate
Pronunciation: KON-DROIT′-EN (DROIT rhymes with Hoyt; EN as in written)

27) Imagine a long molecule of chondroitin sulfate. What types of noncovalent bonds could it form with surrounding molecules?

–Hydrogen (due to OH, C=O, N–H groups)
–Ion-dipole or ion-ion (due to O^- groups)

28) Would you expect chondroitin sulfate to be soluble in water? Explain.

–Yes; the molecule literally bristles with dipolar and ionic charges that would attract water molecules.

Would you expect molecules of chondroitin sulfate to stick together? Explain.

–Yes; but not too tightly; dipolar groups could engage in hydrogen bonding and in ion-dipolar bonding, but the ionic ($-O^-$) groups would repel one another. (or similar response)

(29) List the two major functions of chondroitin sulfate.

–Matrix of cartilage and bone, forms outer coating of cell membrane.

30) What are the two constituents of hyaluronic acid? What are its three major functions?

–Glucuronic acid and N-acetyl glucosamine.
–Intercellular cement, synovial fluid of the joints, vitreous fluid of the eye.
Pronunciation: HI-AL-U-RON′-IC
SIN-O′-VIAL (rhymes with jovial)
VIT′-REE-OUS (VIT- rhymes with fit)

31) What are the constituents of heparin? What is its chief function?

–Glucosamine sulfate and glucuronic acid sulfate.
–As an anticoagulant; particularly in the lining of blood vessels.

(32) What chemical groups are characteristic of mucopolysaccharides?

–Hydroxyl, N-acetyl, sulfate, and carboxyl

(33) List three types of mucopolysaccharides and also list the major functions of each.

–Chondroitin sulfate: matrix of bone and cartilage, outer coat of cell membranes
–Hyaluronic acid: intercellular cement, synovial fluid, vitreous humor of the eye
–Heparin: anticoagulant; especially in linings of the blood vessels

34) The functions of glycoproteins are too numerous to memorize. Do remember that glycoproteins are found on the outer surface of almost all vertebrate cells. What chemical subunit is particularly characteristic of glycoproteins?

–N-acetylneuraminic acid (sialic acid)
Pronunciation: N-ACETYL-NEURA-MIN′-ICK

35) What group is attached to the fifth carbon from the top of N-acetylneuraminic acid?

–N-acetyl ($-NH-CO-CH_3$)

What molecule closely resembles the first 3-carbon structure of N-acetylneuraminic acid?

–Pyruvate

(36) Explain briefly the nature and one chief function of glycoproteins.

–Glycoproteins are molecules in which short polysaccharide chains are attached to proteins. Sialic acid (N-acetylneuraminic acid) is a characteristic subunit of the carbohydrate chain.
–They are attached to the outer surface of all vertebrate cells, among many other functions.

37) What group is attached to the #2 carbon of N-acetylmuramic acid?

–N-acetyl

What molecule does the 3-C group attached to the #3 carbon resemble?

–Lactate
Pronunciation: MURE-AM′-ICK (MURE- rhymes with pure)

(38) Briefly describe the structure of bacterial cell walls.

–The bacterial cell wall is a complex molecular lattice compound of N-acetylmuramic acid, N-acetylglucosamine, and amino acids. (or similar response)

(39) What is the difference between Gram-positive and Gram-negative bacteria?

–Gram-positive bacteria stain with Gram stain.
–Gram-negative do not.
–Gram-negative bacteria have much lipid incorporated in their cell walls (which prevents staining).

(40) Identify each of the following molecules:

A. CH_2OH O OH HO OH OH

B.

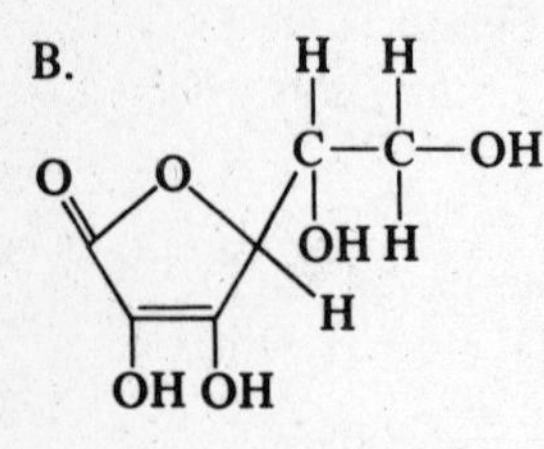

C.

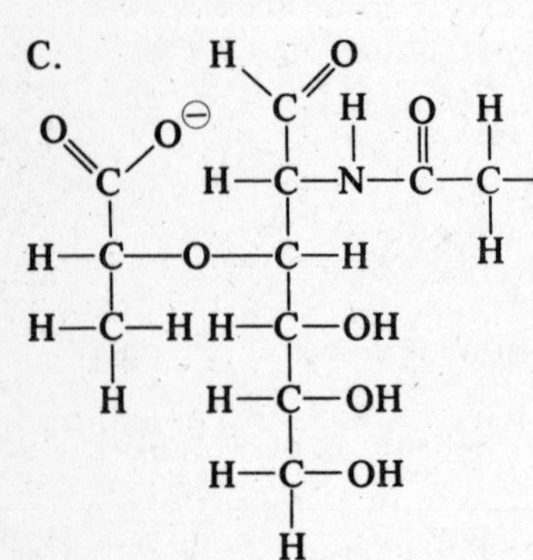

D.

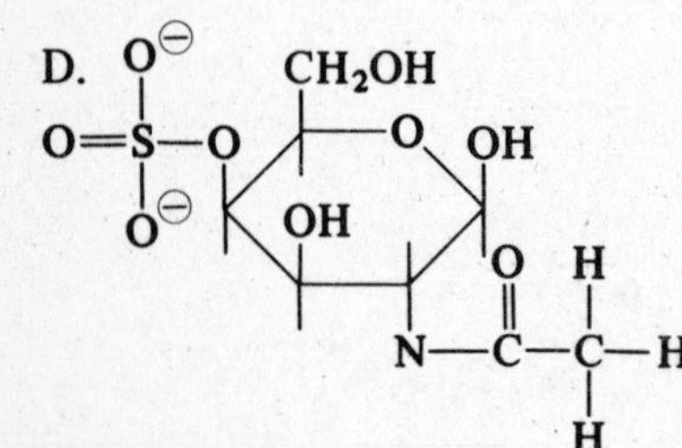

E. O O⊖ C O OH OH HO OH

F. CH_2OH O OH OH

G. CH_2OH O OH OH HO OH

H. O O⁻ C C=O H—C—H H H H—C—OH H—C—C—N—C—H H O HO—C—H H—C—OH H—C—OH H—C—OH H

I. CH_2OH O OH OH OH

J. CH_2OH O OH OH HO NH_2

–A. α glucose
B. Ascorbic acid
C. N-acetylmuramic acid
D. N-acetyl galactosamine sulfate
E. Glucuronic acid
F. Deoxyribose
G. β glucose
H. N-acetylneuraminic acid
I. Ribose
J. β glucosamine

(41) Write the structural formula for an acetyl group and a sulfate group.

O H
‖ |
—C—C—H and —O—S(=O)(O⁻)—O⁻
|
H

Take at least a 5-minute break before going on to the next unit.

SELF-TEST UNITS 9 & 10

(Criterion score 91 points: total of 96 points)

(6 points) 1) What word ending is characteristic of a

_____ salt

_____ sugar

_____ acid

How many carbon atoms does each of the following contain?

_____ triose

_____ nonose

_____ pentose

(3 points) 2) Write the general (empirical) formula for a carbohydrate.

(8 points) 3) Identify each of these molecules by name and functional significance.

A. 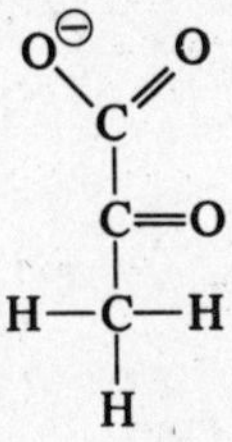 $O^{\ominus}$ O / C / C=O / H—C—H / H

B. H / H—C—OH / C=O / H—C—OH / H

C. O $O^{\ominus}$ / C / H—C—OH / H—C—H / H

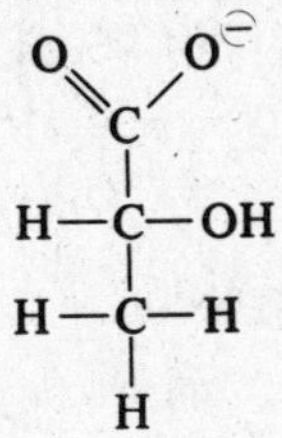

D. 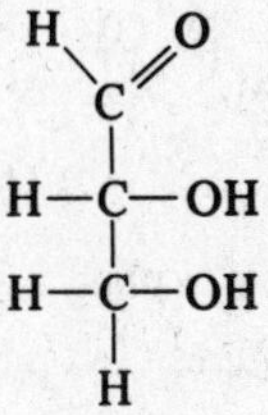 H O / C / H—C—OH / H—C—OH / H

(5 points) 4) Write the structural formula for glucose.

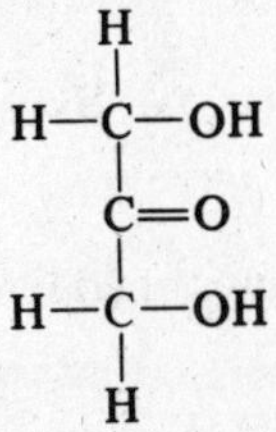

(4 points) 5) Identify each of the following:

A. 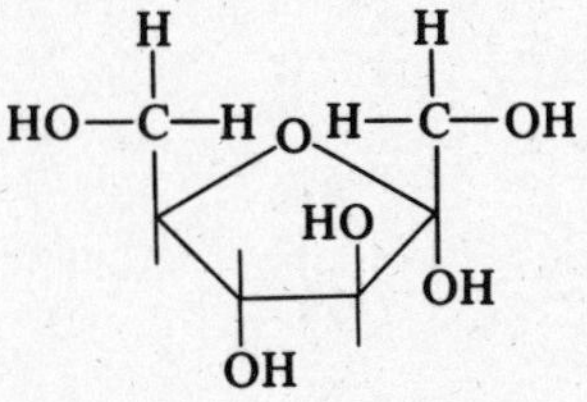 H H / HO—C—H O H—C—OH / HO / OH / OH

C. H / H—C—OH / O / OH / HO OH / OH

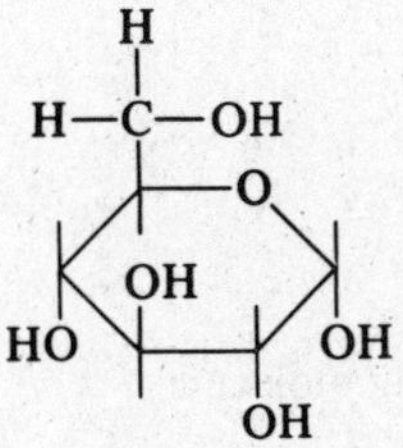

B. 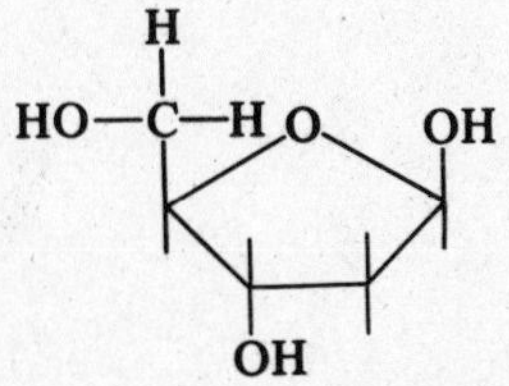 H / HO—C—H O OH / OH

D. 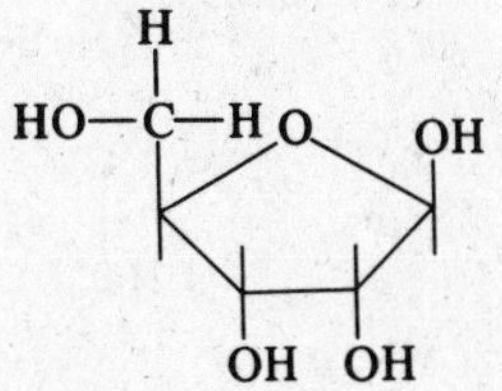 H / HO—C—H O OH / OH OH

(5 points) 6) Diagram the shorthand structural formula for ribose.

(8 points) 7) Label each of the following with the appropriate name and with the appropriate linkage (α or β).

A.

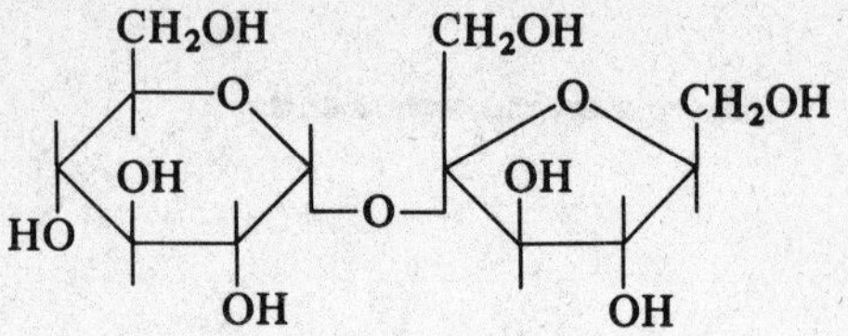

B.

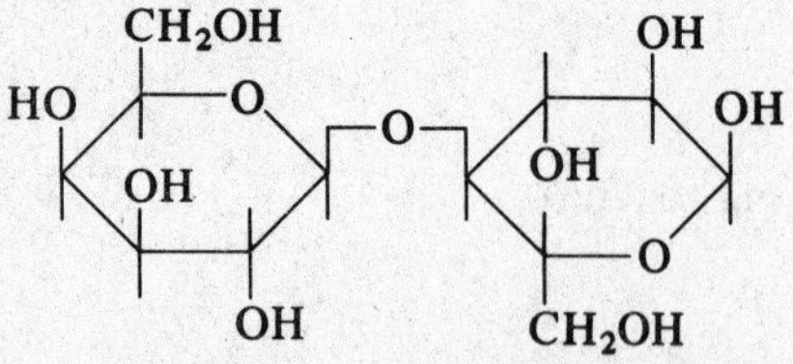

(Galactose) (Glucose)

C.

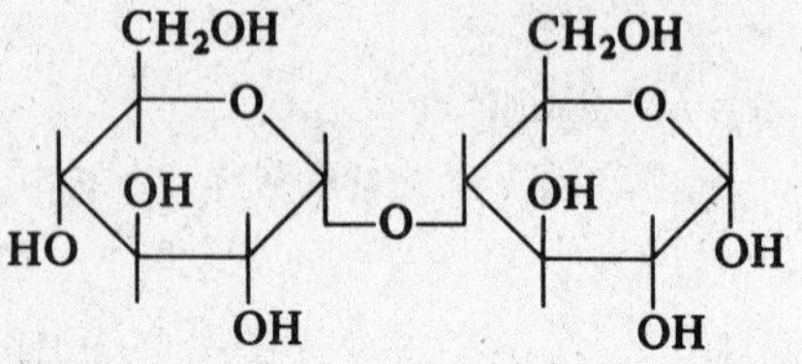

D.

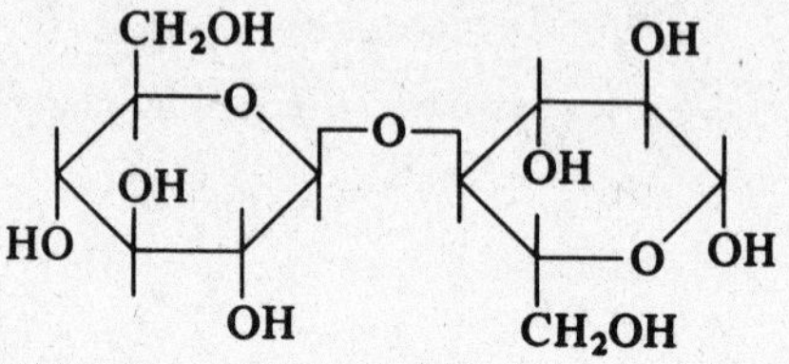

(2 points) 8) Briefly explain a hydrolysis reaction as it applies to disaccharides.

(15 points) 9) Complete the table:

Polysaccharide	Monosaccharide Units	Linkages	Biological Function or Location
Amylose			
Glycogen			
Chitin			
Cellulose			
Amylopectin			

(6 points) 10) Identify each of the following molecules:

A.

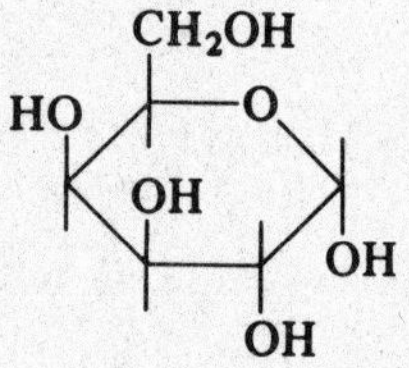

B.

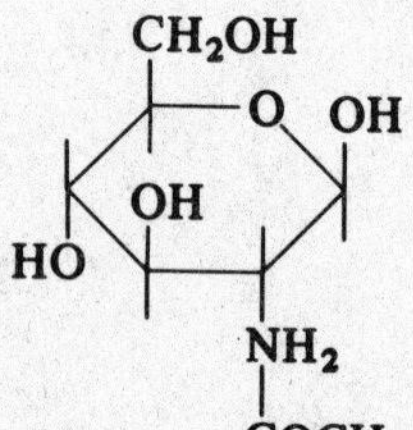

C.

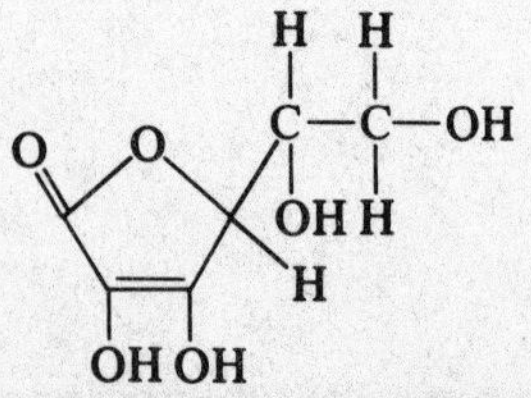

D.

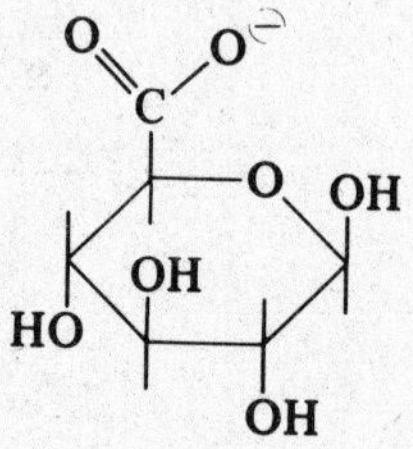

E.

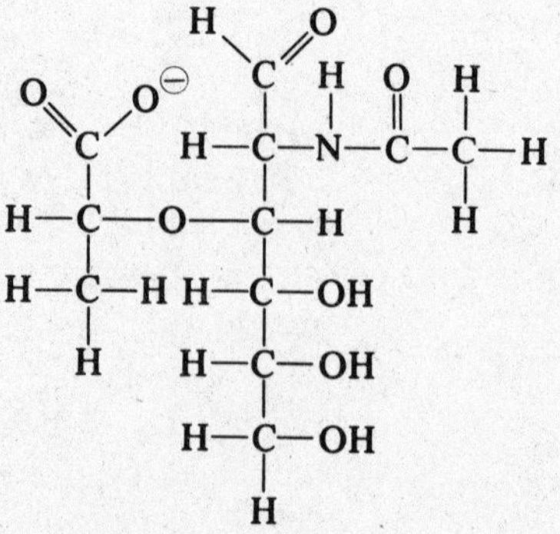

F.

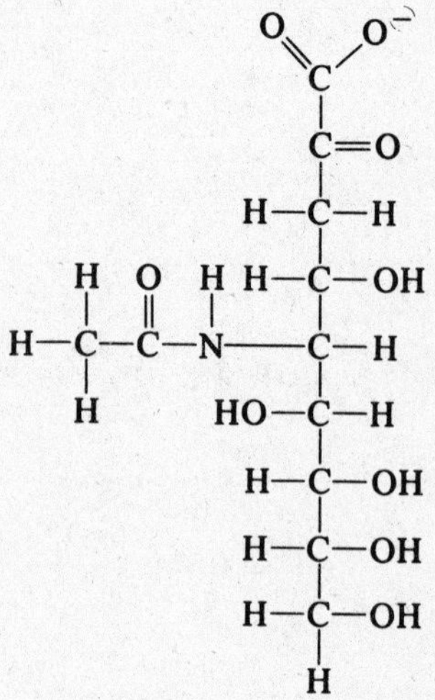

(3 points) 11) List three functions of vitamin C.

(6 points) 12) Describe briefly the nature of mucopolysaccharides, including the types of chemical groups that are characteristically attached to their subunits.

(11 points) 13) List three types of mucopolysaccharides and list the major functions of each.

(4 points) 14) Explain the nature and one major function of glycoproteins.

(4 points) 15) Describe the structure of bacterial cell walls.

(6 points) 16) Write the structural formulae for an acetyl group and for a sulfate group.

(96 points total)

SELF-TEST KEY UNITS 9 & 10

(Criterion score 91 points: total of 96 points)

(6 points) 1) -ate salt — 3 triose
-ose sugar — 9 nonose
-ic acid — 5 pentose
(1 point each) (*Review: Unit 3, Items 16–18; Unit 9, Items 3–6*)

(3 points) 2) $C_n(H_2O)_n$ (3 points) (*Review: Unit 9, Items 1-2*)

(8 points) 3) A. Pyruvate — intermediate in the metabolism of food molecules.
B. Dihydroxyacetone — intermediate in the metabolism of food molecules.
C. Lactate — temporary product in muscle metabolism during exercise.
D. Glyceraldehyde — intermediate in the metabolism of food molecules.
(1 point each) (*Review: Unit 9, Items 8-16*)

(5 points) 4)

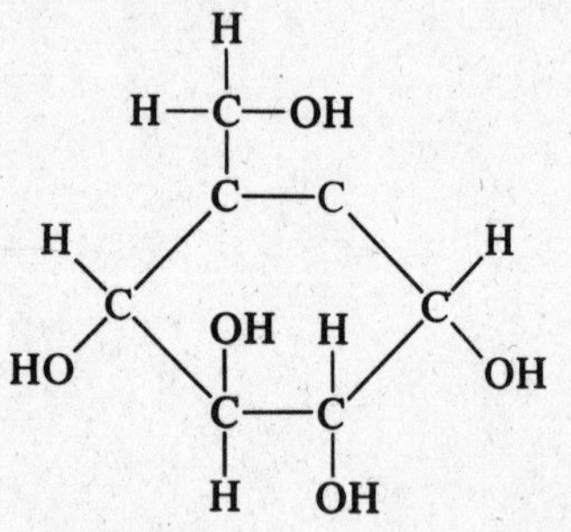

```
            H
            |
         H—C—OH
            |
            C——C
   H      /      \      H
     \   /        \   /
       C            C
     /   \  OH  H  /   \
   HO     \ |   | /     OH
            C——C
            |   |
            H   OH
```

(5 points—one point off for each error) (*Review: Unit 9, Items 17-19*)

(4 points) 5) A. Fructose C. Glucose
B. Deoxyribose D. Ribose
(1 point each) (*Review: Unit 10, Items 1-5
Unit 9, Concepts*)

(5 points) 6)

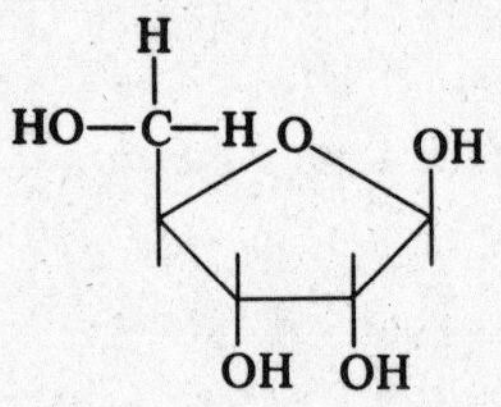

```
       H
       |
  HO—C—H  O       OH
       |  /  \    /
        \/     \ /
        |\     /|
          \___/
          |   |
         OH   OH
```

(5 points—one point off for each error) (*Review: Unit 10, Items 1-2*)

(8 points) 7) A. Sucrose (α) C. Maltose (α)
B. Lactose (β) D. Cellobiose (β)
(1 point each) (*Review: Unit 10, Items 6-11*)

(2 points) 8) A disaccharide may add one molecule of water (1) and split into two monosaccharides (1).
(*Review: Unit 10, Item 4*)

(15 points) 9)

Polysaccharide	Monosaccharide Units	Linkages	Biological Function or Location
Amylose	Glucose	1,4 α	Plant food storage
Glycogen	Glucose	1,4 α and 1,6 α	Animal food storage
Chitin	N-acetyl-glucosamine	β 1,4	Arthropod cuticle
Cellulose	Glucose (β)	β 1,4	Plant cell wall
Amylopectin	Glucose	1,4 α and 1,6 α	Plant food storage

(1 point each) (*Review: Unit 10, Items 13-21*)

(6 points) 10) A. Galactose
B. beta N-acetyl-glucosamine
C. Ascorbic acid
D. Glucuronic acid (glucouonate)
E. N-acetylmuramic acid
F. N-acetylneuraminic acid
(one point each) (*Review: Appropriate concepts pages*)

(3 points) 11) Connective tissue formation (1), wound healing (1), and resistance to disease (1).
(*Review: Unit 10, Items 22-23*)

(6 points) 12) Mucopolysaccharides are gelatinous, sticky, or slimy substances (1) that are polysaccharides composed of subunits that are derivatives of hexoses (1).
Characteristic groups: hydroxyl (1), carboxyl (1), sulfate (1), N-acetyl (1).
(*Review: Unit 10, Items 24-32*)

(11 points) 13)

Chondroitin sulfate (1)	matrix of bone (1) and cartilage (1); outer coat of cell membranes (1).
Hyaluronic acid (1)	intercellular cement (1); synovial fluid of joints (1); vitreous humor of the eye (1).
Heparin (1)	anticoagulant (1); located on inner lining of blood vessels (1).

(*Review: Unit 10, Items 24-33*)

(4 points) 14) Glycoproteins are proteins (1) with attached polysaccharide chains (1). N-acetylneuraminic acid (sialic acid) is a characteristic subunit (1). They are attached to the outer surface of vertebrate cells (1).
(*Review: Unit 10, Items 35-36*)

(4 points) 15) The bacterial cell wall is a complex molecular lattice (1) consisting of N-acetylmuramic acid (1), N-acetylglucosamine (1), and amino acids (1).
(*Review: Unit 10, Items 37-38*)

(6 points) 16) $-\overset{\overset{\displaystyle O}{\|}}{C}-\underset{\underset{\displaystyle H}{|}}{\overset{\overset{\displaystyle H}{|}}{C}}-H \qquad -O-\underset{\underset{\displaystyle O^-}{|}}{\overset{\overset{\displaystyle O^-}{|}}{S}}=O$

(3 points each—one point off for each error)
(*Review: Unit 10, Item 41*)

Take at least a 5-minute break before going on to the next unit.

UNIT 11 ELEVEN AMINO ACIDS

CONCEPTS

Amino acids are the structural units, or building blocks, of proteins. The importance of proteins as both structural and functional elements in living cells cannot be overemphasized. For example, all enzymes are proteins, and nearly all the reactions in a plant or animal organism are made possible by enzyme action. In order to understand the biochemical potential of protein molecules, it is essential to learn to identify and differentiate the amino acids of which they are composed.

General Formula

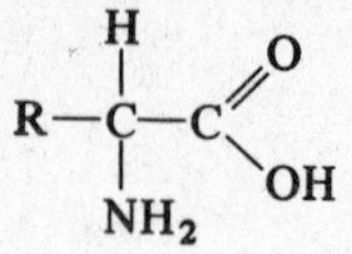

Amino acids are classified by the characteristics of their R groups under physiological conditions. The R group varies with each specific amino acid. The R groups are identified by their shaded backgrounds.

A carboxyl group involves carbon #1. An amino group is attached to carbon #2. The amino acids below are diagrammed in their unionized states.

Table of Amino Acids That Occur In Proteins

Hydrogen Only

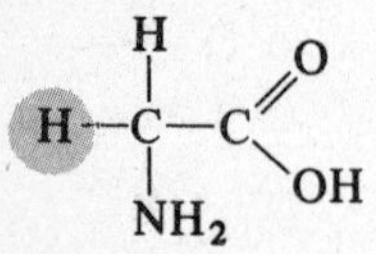

Glycine

Hydroxy-Polar

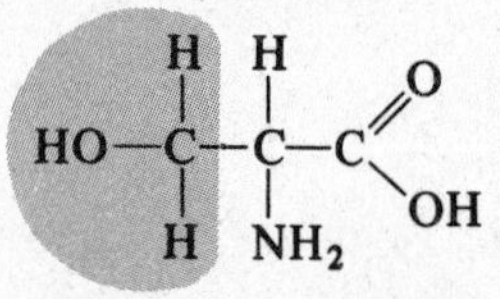

Serine

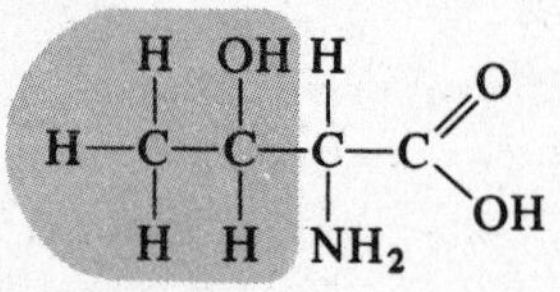

Threonine

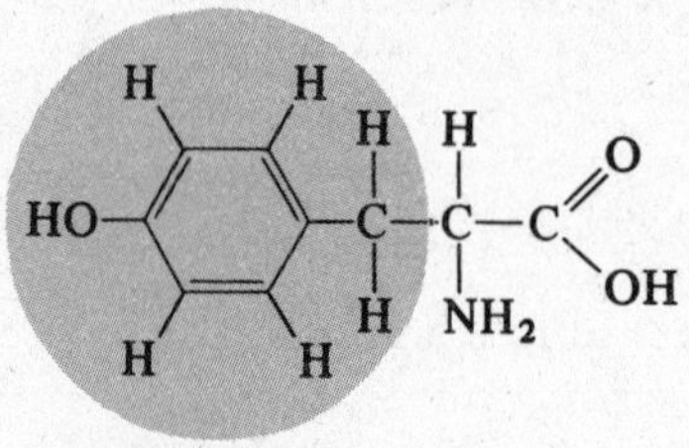

Tyrosine

Ionized-Acidic

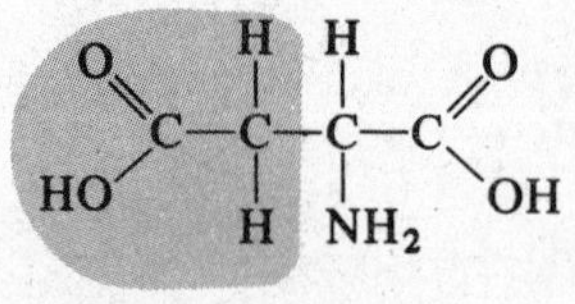

Aspartic Acid

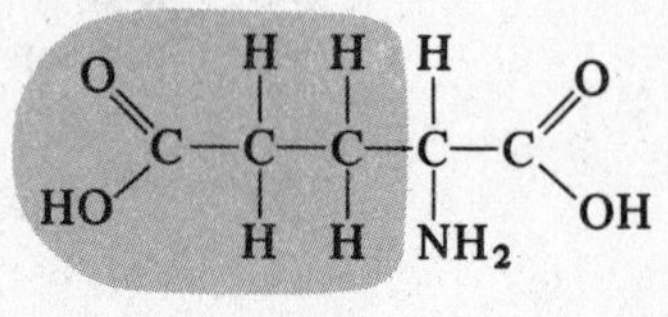

Glutamic Acid

Hydrocarbon

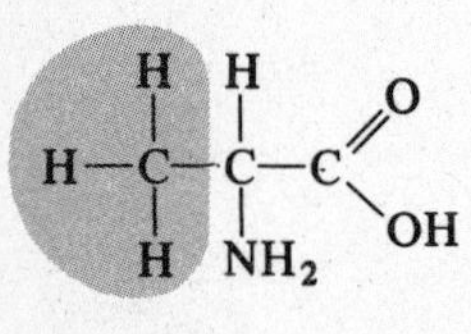

Alanine

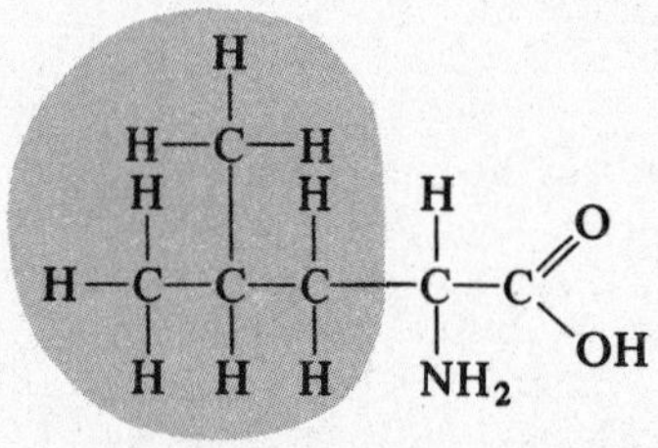

Leucine

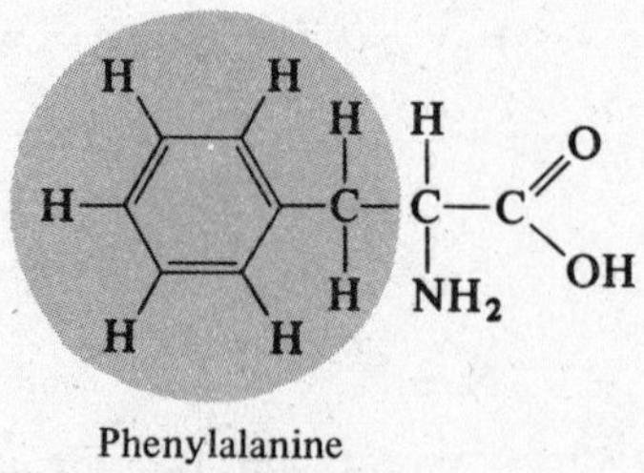

Phenylalanine

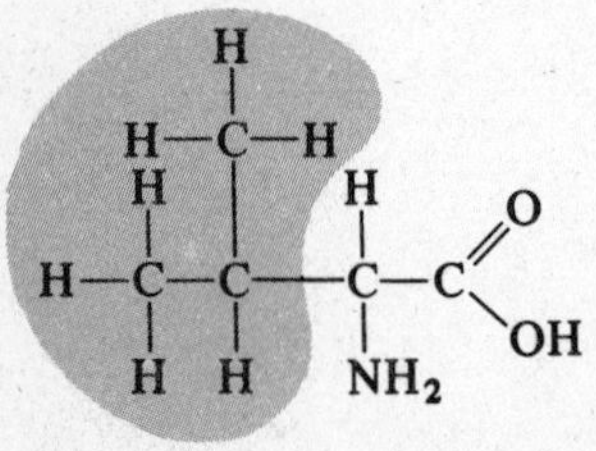

Valine

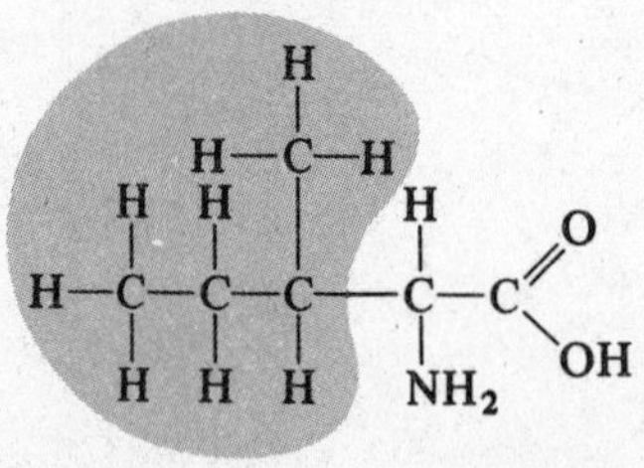

Isoleucine

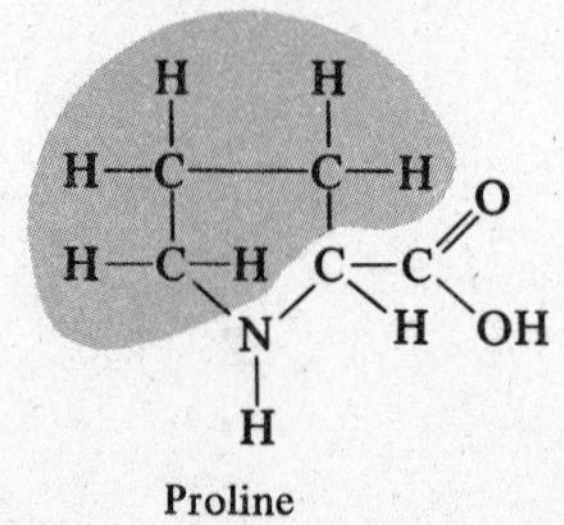

Proline

S-Containing-Polar	N-Containing-Polar	Ionized-Basic

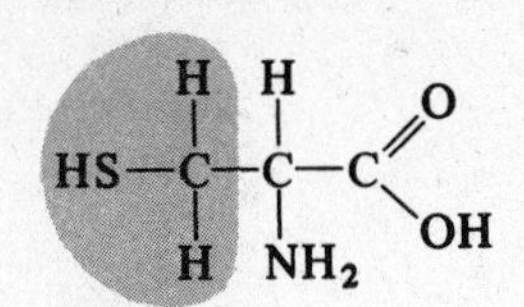

Cysteine

$H_2N-C(=O)-CH_2-CH(NH_2)-C(=O)OH$

Asparagine

$H_2N-CH_2-CH_2-CH_2-CH_2-CH(NH_2)-C(=O)OH$

Lysine

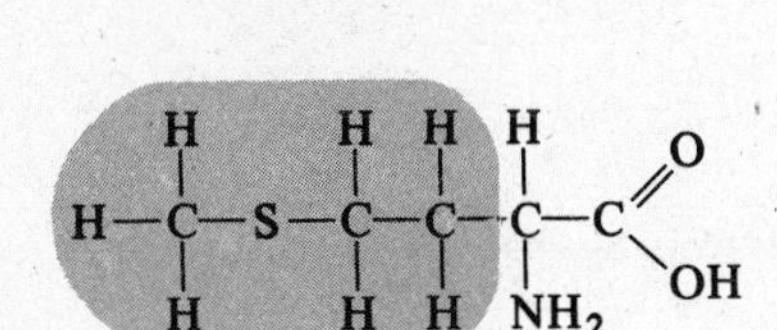

Methionine

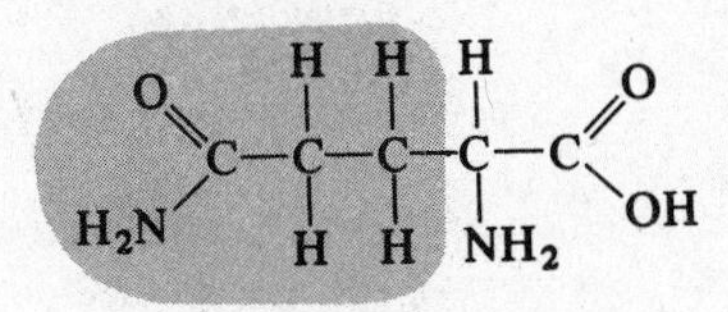

Glutamine

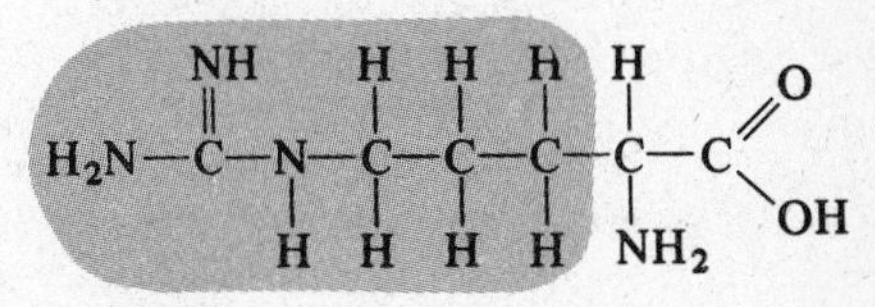

Arginine

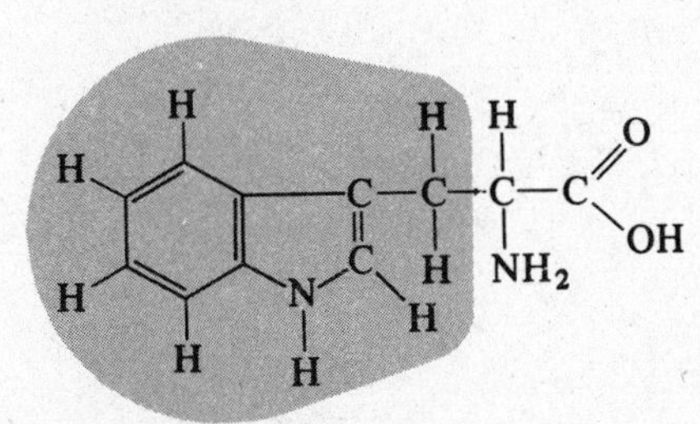

Tryptophan

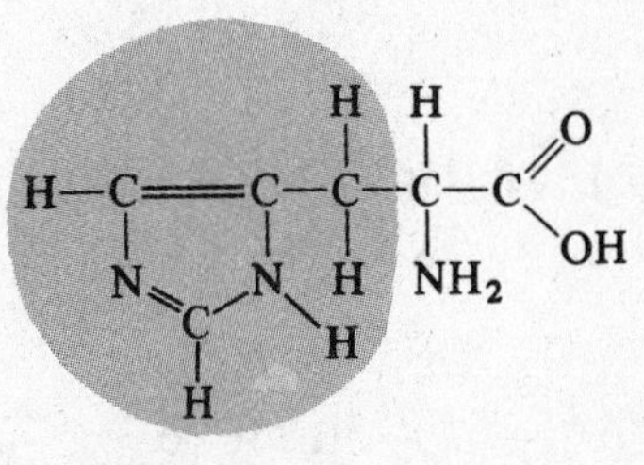

Histidine

QUESTIONS AND PROBLEMS

SAMPLE RESPONSES

1) Write the general structural formula for an amino acid by completing this skeleton:

R–C–C (arrow to the last C: carbon #1)

– $R-CH(NH_2)-C(=O)OH$ or $R-CH(NH_2)-C(=O)OH$ (the second drawing shows the amino group as N with two H atoms)

2) What two chemical groups do all amino acids contain?

–Amino (NH_2)
Acid (COOH or CO_2H)

③ Write the general formula for an amino acid.

– $R-CH(NH_2)-C(=O)OH$

4) Glycine is the amino acid with the simplest structure.

$H-CH(NH_2)-C(=O)OH$

What is the R group of Glycine?

– –H (or H–)
Pronunciation:
GLY′-SEEN (rhymes with sly)

5) Write the structural formula for glycine (without consulting the Concepts).

– $H-CH(NH_2)-C(=O)OH$

6) What do the groups of the hydrocarbon amino acids have in common?

–All 6 R groups contain only C and H atoms.

7) You should learn the names of the six hydrocarbon amino acids and to recognize the structural formula of each. They are usually easiest to remember in this order:

		Pronunciation:
Alanine	(methyl only)	AL′-U-NEEN (U as in up)
Valine	(3 carbons in a V shape)	VAY′-LEAN
Leucine	(4 carbons in a T)	LOU′-SEAN
Isoleucine	(4 carbons in a L)	ISO′-LOU′-SEEN
Phenylalanine	(alanine with a benzine ring attached)	FENNEL′-AL′-U-NEEN -ine is always pronounced -een.
Proline		PRO′-LEAN

Say each word *out loud* three times.

What is peculiar about the #2 C and the amino group of proline?

–The #2 C and the amino group are part of the five member ring structure.

8) Write the correct name for each of these amino acids:

A.

B.

C.

D.

E.

F.

G.

–A. Proline
B. Alanine
C. Isoleucine
D. Glycine
E. Valine
F. Phenylalanine
G. Leucine

9) List the six hydrocarbon amino acids.

–Alanine, valine, leucine, isoleucine, phenylalanine, and proline.

(10) What effect would these six hydrocarbon R groups have on the solubility in water of proteins that contain them?

–They would retard solubility in water.
They are hydrophobic.
(or similar response)

11) What characteristic is common to the R groups of the hydroxy-polar amino acids?

–Each R group contains an –OH group (which is not in a carboxyl group).

Pronunciation:
SEAR′-EEN
THREE′-U-NEEN (U as in up)
TIE′-ROW-SEEN
Repeat each name *out loud* three times.

12) What effect would these hydroxy-polar R groups have on the solubility of the proteins that contain them?

–They would facilitate solubility in water because they are polar.

13) How do the structural formulae of serine and threonine differ?

–If one of the H's in the R group of serine is replaced by a methyl group, threonine is produced.

14) What other amino acid is a close relative of tyrosine? Explain.

–Phenylalanine
–The terminal H in the benzine ring of phenylalanine is replaced by an –OH group in tyrosine.

15) List the three hydroxypolar amino acids.

–Serine, threonine, and tyrosine

16) Describe the R groups of the two S-containing-polar amino acids. An –SH group is called a sulfhydril group.

–Cysteine: CH_2 plus a sulfhydril group.
–Methionine: CH_2–CH_2 plus S plus a methyl group.
Pronunciation:
CYST′-E-EEN
METH-EYE′-O-NEEN (METH- as in methyl)
Say each word *out loud* three times.

17) What effect would these S-containing polar groups have on the solubility in water of a protein that contains them?

–They would promote solubility in water because they are polar.

18) Two cysteine molecules can combine by linking their –SH groups, forming a –S–S– disulfide bond and releasing 2 H atoms. The resulting molecule is a double amino acid called cystine.

Using the structural formula for cysteine from the Concepts page, diagram the structural formula for cystine.

```
            H   H         H   H
   O        |   |         |   |        O
    \\      |   |         |   |      //
–     C — C — C — S — S — C — C — C
    /       |   |         |   |      \
  HO       H2N  H         H  NH2      OH
```

19) List the names of the two sulfur containing polar amino acids.

–Cysteine and methionine.

20) Write the correct name for each of these molecules:

```
A.     H       H  H  H
       |       |  |  |     O
   H—C—S—C—C—C—C//
       |       |  |  |     \
       H       H  H  NH2    OH

                       H H
B.       H  H
     O   |  |
      \\C—C—C—[ring]—OH
     HO/  |  |
        H2N  H
                       H H

C.    O    H  H            D.        H  NH2
       \\  |  |                      |  |     O
         C—C—C—OH               HS—C—C—C//
     HO/   |  |                      |  |    \
         H2N  H                      H  H     OH

E.    O    H2N H  H
       \\  |   |  |
         C—C—C—C—H
     HO/   |   |  |
           H  OH  H
```

—A. Methionine
B. Tyrosine
C. Serine
D. Cysteine
E. Threonine

21) What characteristic do each of the three N-containing polar amino acids have in common?

—They all contain N atoms in the R groups; hence the R groups are polar. (The =O groups of asparagine and glutamine also add to the polarity of their R groups.)

—Pronunciation:

ASS-PARE′-U-JEAN (U as in up)
GLUE′-TU-MEAN (U in TU as in up)
TRIP′-TOE-FAIN (rhymes with pain)

(22) What effect would these N-containing R groups have on the solubility in water of the proteins that contain them?

—These N-containing R groups would facilitate solubility in water.

23) Name the three N-containing-polar amino acids.

—Asparagine, glutamine, and tryptophan.

24) How many rings does tryptophan's R group contain?

—Two

How many CH_2 groups do asparagine and glutamine each contain?

—Asparagine: 1; glutamine: 2

25) What characteristic is common to the R groups of the acidic amino acids?

—The R groups both contain a carboxyl group (organic acid).

—Pronunciation:

ASS-PAR′-TICK
GLUE-TAM′-ICK

26) How does asparagine differ from aspartic acid and how does glutamine differ from glutamic acid?

—In both cases a $-NH_2$ group replaces the $-OH$ group of the carboxyl group of the R group.

(27) Would you suspect that the carboxyl group of aspartic and glutamic acids' R groups are dissociated at physiological pH? Explain briefly.

—Yes. All carboxyl groups are dissociated at pH = about 7.2. Hence, *all* the carboxyl groups of *all* amino acids are dissociated at physiological conditions and bear ⊖ charges.

(28) What effect would these ⊖ charged R groups have on the solubility in water of proteins that contain them?

–They would increase solubility in water due to ion-dipole bonding with water molecules.

29) Name the two acidic amino acids.

–Aspartic acid and glutamic acid

30) Groups that bind H ions (H^+) are called *basic* groups.

An amino group ($-NH_2$) is commonly a basic group at physiological pH. Below a specific pH the N⊖ of an amino group attracts and binds an H^+ and the amino group is then $-NH_3^{\oplus}$.

–Pronunciation:
LIE′-SEEN
ARE′-GIN-EEN
HISS′-TU-DEAN (U in TU as in up)
Repeat each word three times out loud.

What characteristic is common to the R groups of lysine and arginine?

–The R groups both contain an amino group.
–Above specific H^+ concentrations (below specific pH values) this amino group will bind a H^+ and become $-NH_3^{\oplus}$.

(31) Will the amino group on all amino acids be charged at physiological pH?

–Yes

32) Under physiological conditions, the R groups of histidines that are incorporated into protein molecules are commonly + charged. The =N⊖ of histidine's R group may bind a H^+ and therefore act as a basic group.

Diagram the =N⊖ of histidine's R group after it has bound (associated with) a H^+.

–=$N^{\oplus}$–

(33) Will the basic R groups of the basic amino acids affect the solubility in water of proteins that contain them?

–Yes, the basic R groups will facilitate solubility in water due to ion-dipole bonding with water molecules.

34) Name the three basic amino acids.

–Lysine, arginine, and histidine.

35) Identify each of these amino acids by name.

A.
```
O   OH
 \\ /
  C
  |
H—C—NH2
  |
H—C—H
  |
  C
 // \
O    OH
```

B.
```
O   OH
 \\ /
  C
  |
H—C—NH2
  |
H—C—H
  |
H—C—H
  |
H—C—H
  |
H—C—H
  |
  NH2
```

C.
```
O   OH
 \\ /
  C
  |
H—C—NH2
  |
H—C—H
  |
H—C—H
  |
  C
 // \
O    NH2
```

D.
```
             H  H
             |  |       O
H—C====C—C—C—C//
  |    |  |  |  \
  N    N  H  NH2  OH
   \\ / \
     C   H
     |
     H
```

E.
```
     NH    H  H  H  H
     ||    |  |  |  |      O
H2N—C—N—C—C—C—C—C//
        |  |  |  |  |  \
        H  H  H  H  NH2 OH
```

–A. Aspartic acid
B. Lysine
C. Glutamine
D. Histidine
E. Arginine

F.

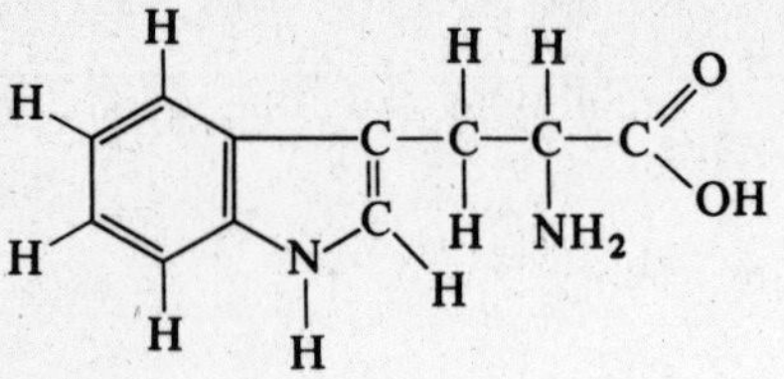

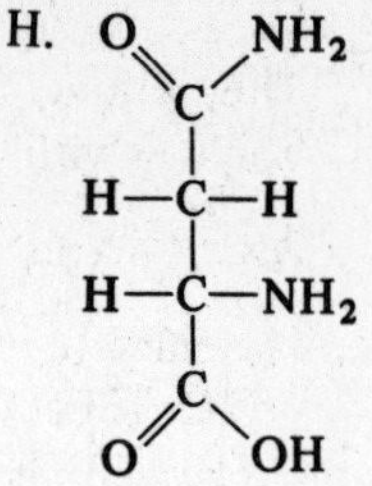

G.

H.

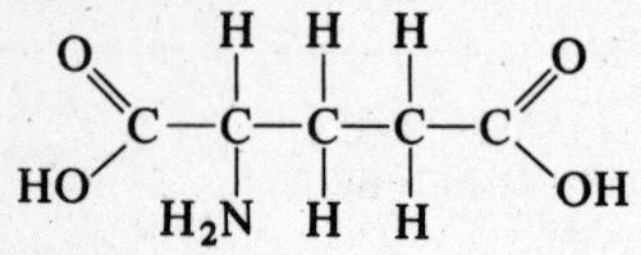

–F. Tryptophan
G. Glutamic acid
H. Asparagine

(36) List the seven classes of amino acids

–Hydrogen only, hydrocarbon, hydroxy-polar, S-containing polar, N-containing polar, ionized-acidic, and ionized-basic.

(37) Name the N-containing polar amino acids.

–Asparagine, glutamine, and tryptophan.

(38) Name the H-only amino acid and list the hydrocarbon amino acids.

–Glycine
–Alanine, valine, leucine, isoleucine, phenylalanine, and proline

(39) List the S-containing polar amino acis.

–Cysteine and methionine

(40) List the basic amino acids.

–Lysine, arginine, and histidine

(41) List the acidic amino acids.

–Aspartic acid and glutamic acid

(42) List the hydroxypolar amino acids.

–Serine, threonine, and tyrosine

(43) Identify each amino acid by name:

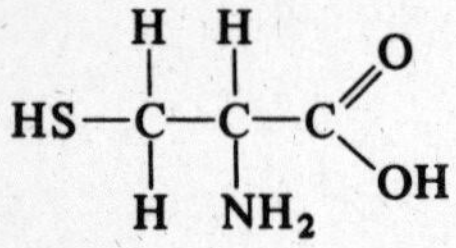

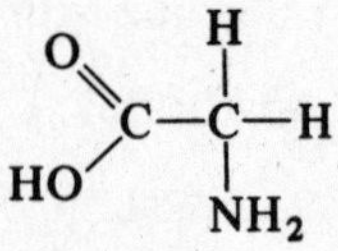

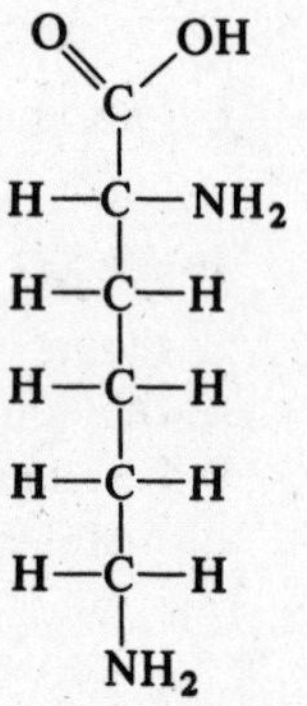

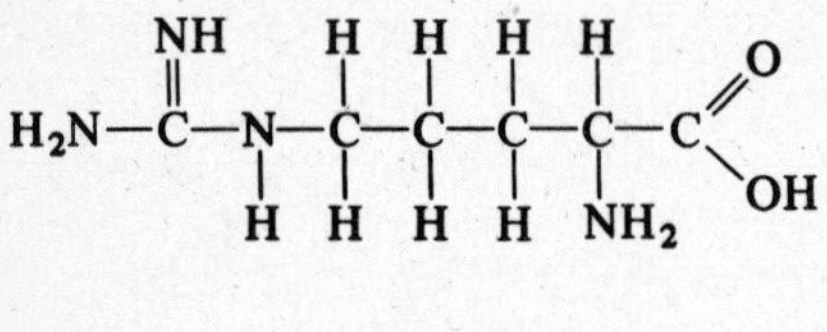

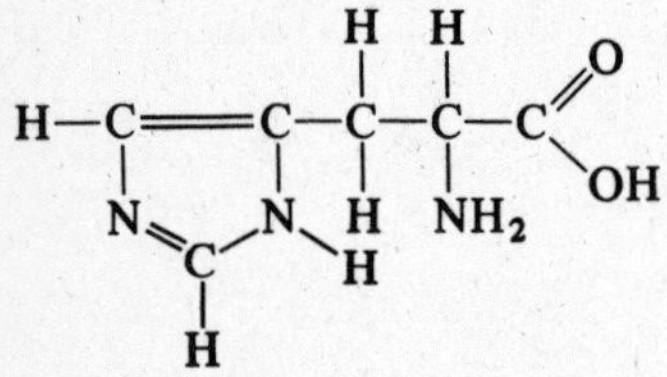

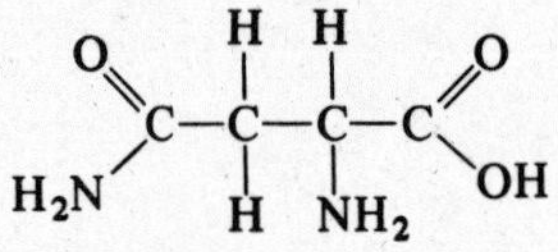

–Cysteine Glycine Lysine

Arginine

Histidine Asparagine

(44) Identify each amino acid:

```
      H  H    O             O     H  H
      |  |   //              \\   |  |
  HO—C——C——C                 C——C——C—H
      |  |   \              /   |  |
      H  NH2  OH          HO   H2N H

  H       H  H  H                O     OH
  |       |  |  |    O            \\  /
H—C—S——C——C——C——C//               C
  |       |  |  |   \             |
  H       H  H  NH2  OH        H—C—NH2
                                  |
                               H—C—H
      H  H                   H      H
 O    |  |    O              (benzene ring)
  \\  |  |   //              H      H
    C—C——C——C                   H
  /   |  |   \
HO   H2N H    OH

         H                  H         H
         |                  |         |
      H—C—H              H—C————————C—H
  H  H   |   H                |         |      O
  |  |   |   |    O       H—C—H      C——C//
H—C—C——C——C——C//              \      /  \
  |  |   |   |    \             N     H   OH
  H  H   H   NH2   OH           |
                                H
```

–Serine

Methionine

Aspartic acid

Isoleucine

Alanine

Phenylalanine

Proline

(45) Identify each amino acid:

```
   H      H                        H  H  H
          H  H    O         O      |  |  |
HO—(benzene)—C——C——C//       \\    |  |  |
          |  |    \           C——C——C——C—H
          H  NH2   OH       /    |  |  |
   H      H               HO   H2N OH H

      H                           H
      |                           |
   H—C—H                       H—C—H
   H  |   H  H                 H  |   H
   |  |   |  |    O            |  |   |    O
H—C——C——C——C——C//           H—C——C——C——C//
   |  |   |  |    \            |  |   |    \
   H  H   H  NH2   OH          H  H   NH2   OH

O      H  H          H  H     O
 \\    |  |          |  |    //
   C——C——C——S——S——C——C——C
  /    |  |          |  |    \
HO   H2N  H          H  NH2   OH

    H           H  H
H                |  |    O
   (ring)   C——C——C——C//
               ||     |   \
H          N—C   H  NH2  OH
    H       |   H
            H

O     H  H  H     O        O     H  H  H     O
 \\   |  |  |    //         \\   |  |  |    //
   C——C——C——C——C              C——C——C——C——C
  /   |  |  |    \          /    |  |  |    \
HO    H  H  NH2   OH      H2N    H  H  NH2   OH
```

–Tyrosine

Leucine

Cystine

Tryptophan

Glutamic acid

Threonine

Valine

Glutamine

(46) Identify the group (category) to which each of the amino acids belongs:

Alanine	Leucine	–Hydrocarbon	Hydrocarbon
Arginine	Lysine	Basic	Basic
Asparagine	Methionine	N-polar	S-polar
Aspartic acid	Phenylalanine	Acidic	Hydrocarbon
Cysteine	Proline	S-polar	Hydrocarbon
Glutamic acid	Serine	Acidic	OH-polar
Glutamine	Threonine	N-polar	OH-polar
Glycine	Tryptophan	H-only	N-polar
Histidine	Tyrosine	Basic	OH-polar
Isoleucine	Valine	Hydrocarbon	Hydrocarbon

Take at least a 5-minute break before continuing with the next unit.

UNIT TWELVE 12

PROTEINS I: PEPTIDE BONDING AND POLYPEPTIDES

CONCEPTS

Many proteins are important in biological structure—for example, the keratin of hair, collagen of skin and leather, and fibroin of silk. Other proteins function as enzymes, which control or modulate literally all biochemical reactions. The biochemical potential of a protein is determined by its constituent amino acids.

The actual synthesis of proteins takes place on the ribosomes and also involves messenger RNA (m-RNA), transfer RNA (t-RNA), ATP, GTP, enzymes, and other proteins.

A protein is a chain of amino acids that are linked together by peptide bonds; each amino acid forms one peptide unit in the protein molecule. For example:

Three amino acids

$$3\ H_2N-CHR-COOH \longrightarrow H_2N-CHR-\boxed{CO-NH}-CHR-\boxed{CO-NH}-CHR-COOH + 2\ H_2O$$

Tripeptide

Peptide bonds

The sequence of amino acids in a protein is usually indicated by the use of standard abbreviations:

ala-arg-asn-asp-cys-glu-gln-gly-his-ile-leu-lys-met-phe-pro-ser-thr-trp-tyr-val

Primary Structure

The primary structure of a polypeptide involves the sequence of amino acids in the chain. Primary structure is genetically coded by DNA.

Below is the primary structure of beef insulin, a protein hormone of the pancreas that regulates sugar metabolism.

Ⓐ S—S

gly-ile-val-glu-gln-cys-cys-ala-ser-val-cys-ser-leu-tyr-gln-leu-glu-asp-tyr-cys-asn

Ⓑ S—S Ⓒ S—S

phe-val-asn-gln-his-leu-cys-gly-ser-his-leu-val-glu-ala-leu-tyr-leu-val-cys-gly-glu-arg-gly-phe-phe-tyr-thr-pro-lys-ala

QUESTIONS AND PROBLEMS

1) Circle any of the following that are *not* amino acids.

$R-CH(NH_2)-COOH$ (H above, NH_2 below)

$R-CH(NH_2)-COOH$ (NH_2 above, H below)

$H-CR(NH_2)-COOH$ (NH_2 above, R below)

$H-CR(NH_2)-COOH$ (R above, NH_2 below)

$H-CR(NH_2)-COOH$ (R above, N with H, H below)

SAMPLE RESPONSES

—All are amino acids.

2) Circle the H and OH that will be removed as two amino acid molecules are linked. What two *atoms* will then be linked by a covalent bond? What two *groups* are so linked?

$R{-}CH(NH_2){-}C(=O)OH \quad H_2N{-}CH(R){-}C(=O)OH$

— $R{-}CH(NH_2){-}C(=O)$(OH) (H)$N(H){-}CH(R){-}C(=O)OH$ [OH and H circled]

—C and N
—Carbonyl (acid) and amino

3) Complete the reaction (you may consult the preceding item if necessary):

$R{-}CH(NH_2){-}C(=O)OH + R{-}CH(NH_2){-}C(=O)OH \rightleftharpoons$ +

— $R{-}CH(NH_2){-}C(=O){-}NH{-}CH(R){-}C(=O)OH + H_2O$

4) Each amino acid in this structure is a *peptide*. Circle each peptide unit in the *dipeptide* molecule below. What two atoms are joined in the peptide bond?

$R{-}CH(NH_2){-}C(=O){-}NH{-}CH(R){-}C(=O)OH$

peptide bond

— ($R{-}CH(NH_2){-}C(=O)$) ($NH{-}CH(R){-}C(=O)OH$)

—C and N

5) How many amino acid molecules would be involved in the formation of the following:

Dipeptide________
Tripeptide_______
Pentapeptide_____
Octapeptide______
Polypeptide______

—Dipeptide 2
Tripeptide 3
Pentapeptide 5
Octapeptide 8
Polypeptide many

⑥ Write the equation for the formation of a peptide bond between two amino acids of general structural formulae. Label the peptide bond.

— $H{-}CR(NH_2){-}C(=O)OH + H_2N{-}CH(R){-}C(=O)OH \rightleftharpoons$

$H{-}CR(NH_2){-}C(=O){-}NH{-}CH(R){-}C(=O)OH + H_2O$

Peptide bond

7) Write the general structural formula for a pentapeptide. Circle each peptide unit.

— 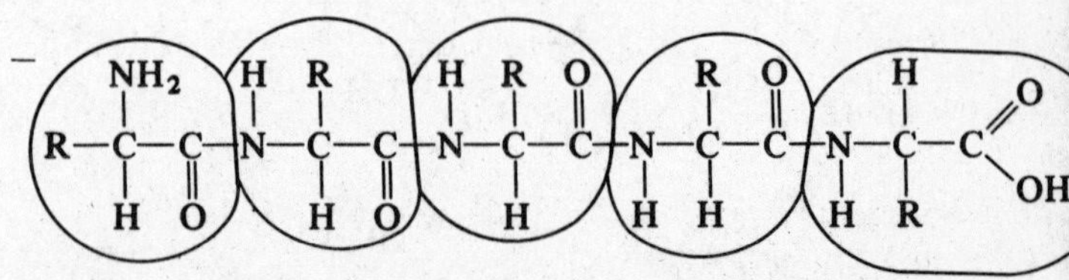

The R, =O, and H (attached to N) may be oriented either up or down, but the =O and adjacent N's H should be on opposite sides of the backbone.

8) What is the repeating sequence of atoms in the backbone of a polypeptide? (Consult your last answer.)

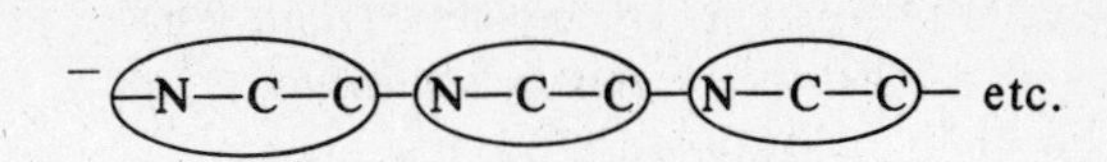

9) For convenience, we might diagram a polypeptide thus:

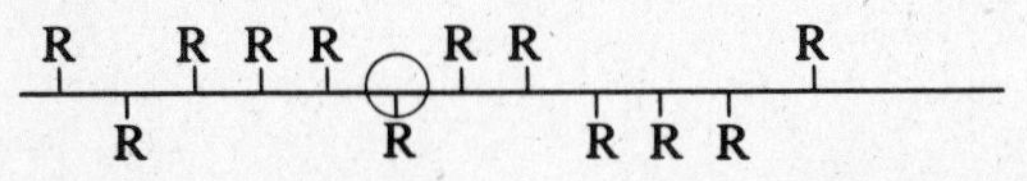

Each R = the R group of one amino acid. How many peptides are involved in this structure? What atoms would the area included within the circle represent? (Use a structural formula.)

–12

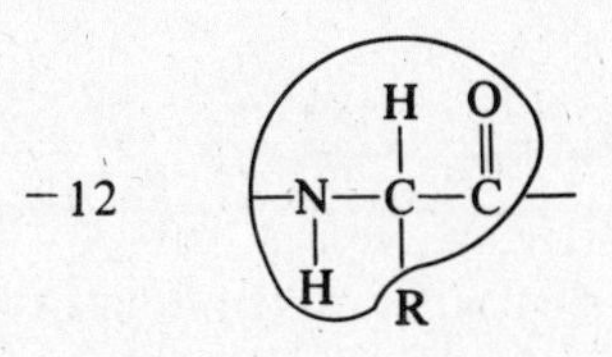

10)

Cysteine (example)

A. B. C. D. E. F.

In this heptapeptide, label each peptide unit with its appropriate amino acid name.

–A. Glycine
B. Alanine
C. Aspartic acid
D. Lysine
E. Serine
F. Phenylalanine

(11) A *protein* molecule is a large polypeptide. The size varies from about 50 peptide units to thousands, with several hundred peptides the most common number. Which of the following *could* be proteins? (Assume 1 peptide = M.W. of 100 daltons.)

Molecular Weight
96
256
1,420
8,460
22,400
465,000

–

96	No
256	No
1,420	No (14± peptide units)
8,460	Yes (84± peptide units)
22,400	Yes (224± peptide units)
465,000	Yes (4650± peptide units)

12) What is the approximate molecular weight range of proteins? What are the "building blocks" of proteins?

–5,000 to thousands (or approximate equivalent) of amino acids (peptides)

13) Name the first 7 peptide units of the longer chain of beef insulin. (Use the diagram in the Concepts.)

–Phenylalanine, valine, asparagine, glutamine, histidine, leucine, and cysteine

14) What type of bonds holds the two chains of beef insulin together?

–Disulfide bonds (–S–S–) between cysteine units (forming cystine)

15) In the diagram of the insulin molecule, which disulfide bonds (A, B, or C) are intrachain bonds? (Hint: What is the difference between *intra*murals and *inter*collegiate athletics?) Which disulfide bonds are interchain bonds?
Is insulin composed of one or two polypeptides?

–Intrachain: A
–Interchain: B and C
–Two

16) Think back to the definitions of bonding. Can a bond really be as long as S–S bond "A"?
In the insulin diagram, if the two cysteines are joined by disulfide bond "A," is the chain between them folded or straight? How many terminal $-NH_2$ groups does insulin have? How many terminal –COOH groups does insulin have?

–No

–Folded
–Two
–Two

17) How many terminal $-NH_2$ and –COOH groups would a tripeptide have?
. . . beef insulin?

–One $-NH_2$ and one –COOH (at each end of the chain)
–Two $-NH_2$ and two –COOH (due to two chains)

18) Several peptide units are ionically charged at the normal pH of cells and body fluids:

⊕ Basic R groups of lysine, arginine, and histidine. The terminal $-NH_2$ of the polypeptide chain.

⊖ Acidic R groups of aspartic and glutamic acids. The terminal –COOH of the polypeptide chain.

Memorize these 5 R groups and two terminal groups.

Which peptide units of the smaller chain of insulin will be charged in the human body? (The terminal $-NH_2$ end is to the left.) Give the charge for each.

– 1 glycine + (terminal $-NH_3^{\oplus}$)
4 glutamic acid -
17 glutamic acid -
21 asparagine - (terminal $-COO^{\ominus}$)

(19) Which peptide units of the longer subchain of insulin are normally charged?

– 1 phenylalanine +
5 histidine +
10 histidine +
13 glutamic acid -
21 glutamic acid -
22 arginine +
29 lysine +
30 alanine -

20) Most proteins have more - charges than + charges on their peptide units. If a typical protein is placed in solution in an electrophoretic apparatus with + and - electrodes connected to a D.C. electrical power supply, toward which electrode will it migrate?

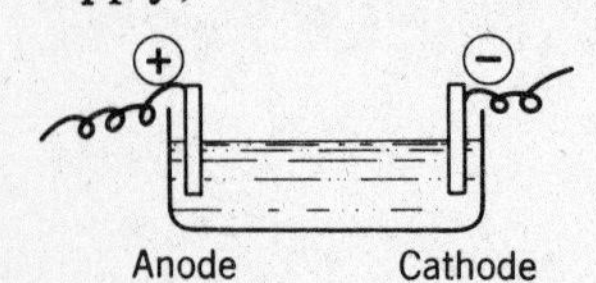

–Toward the anode (+ electrode)

21) The migration of molecules in an electrical field is called electrophoresis.

Toward which electrode of an electrophoretic apparatus would beef insulin migrate?

–Pronunciation: electro-FORE-E′-SIS

–Insulin would probably not migrate, since the total + charges = total - charges. (6 + = 6 -)

(22) The + charge on histidine is very sensitive to changes in the pH of the medium. If the pH drops just below the physiological range, histidine loses its H^+ and its charged group becomes =N:.

Toward which electrode would insulin migrate in a solution of pH = 5?

–Insulin would now migrate toward the anode. (6 – and 4 +)
(Dipolar charges ⊕ and ⊖ do not influence the direction of electrophoretic migration.)

23) Cut out the α-helix model (α = alpha) that is located opposite page 96. Note that this model does not contain all the atoms in a polypeptide chain. Diagram a short segment of a polypeptide chain and circle the atoms that are omitted in the model.

```
   (H)       H (H)       H
    |        |  |        |
 --C---C---N---C---C---N--
    |   ||      |   ||
    R   O       R   O
```

24) Fold out (up) the R groups on the α-helix model. Which of the remaining atoms in the polypeptide chain can participate in H bonding?

–The H in –N–H
The O in C=O

25) *From the right*, locate the *first* O and the *fifth* H. Bend the left end of the model down and to the right to form a spiral (helix)–matching the fifth H to the first O, thus

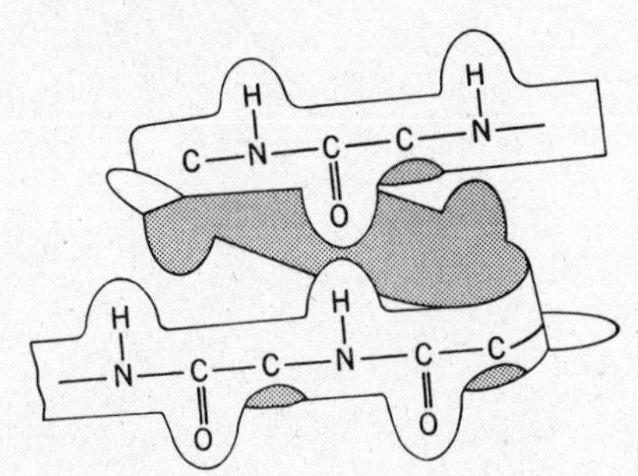

What type of bond can form here?

–H bond

26) Grasp this H bond (the =O *and* H) between the thumb and forefinger of your right hand. Continue to spiral the model, matching each $\overset{||}{O}$ with each H to form H bonds. How many H bonds (including the one under your left thumb) does this polypeptide chain form?

–6 (There are 3.6 peptides per "turn" of the helix)

This structure for a protein, the α helix, was first postulated by Dr. *Linus Pauling* in 1951. He received the Nobel Prize in Chemistry in 1954 for this brilliant achievement. (*Remember Pauling's name.*) Note that the R groups (bent out) literally "bristle out" from all sides of the α helix. What would you predict as to the ability of an α-helical protein to react with adjacent molecules.

–Great reactivity due to R groups.

(27) What term is applied to the polypeptide configuration you have just been dealing with? This structure is characteristic of some proteins but not all proteins. Who first set forth the idea of this structure?

–α helix
–Dr. Linus Pauling

(28) In your own words, explain what an alpha helix is.

–An alpha helix is the regular coiling of a polypeptide chain.
–It is stabilized by H bonding between the $=O^{\ominus}$ and $-NH^{\oplus}-$ of peptide units.
–There are 3.6 peptide units per "turn" of the helix. (or similar response)

29) Formation of the alpha helix requires that the –NH–CHR–CO– backbone of each peptide unit be sufficiently flexible to twist into the helical configuration. Proline prevents alpha-helix formation in the region of a protein in which it occurs. Explain why you think that proline cannot participate in alpha-helix coiling.

–The N of proline is a part of the ring structure; hence it is not flexible enough to form the coil of the helix.
–The –N– of proline does not have a H attached when it is in a polypeptide chain; hence it cannot participate in the H bonding of the alpha helix.

(30) Which amino acid is commonly found at the ends of alpha helices in proteins? It is referred to as "terminating" the alpha-helical portions of the protein.

–Proline

Take at least a 5-minute break before going on to the next unit.

α-helix model. Cut out along line

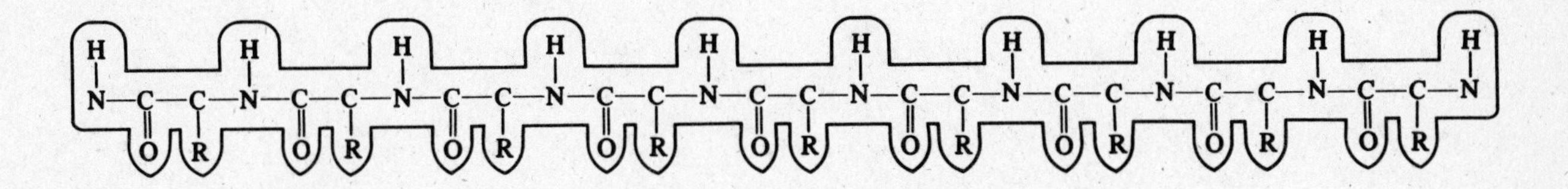

UNIT THIRTEEN PROTEINS II: LEVELS OF STRUCTURE AND CONJUGATED PROTEINS 13

CONCEPTS

The structure of a protein molecule is extremely important in determining its biological activity. Biochemists recognize four levels or types of structure within a protein. These levels of structure are designated very simply: primary, secondary, tertiary, and quaternary. You have already dealt with the first two levels of structure as you formed peptide bonds and constructed the alpha helix.

Conformation is a general term for the twisting or folding of the polypeptide chain.

Primary structure involves the sequence of amino acids in the polypeptide chain.

Secondary structure is the formation of alpha helices and other recurring patterns along one or two dimensions (e.g., the beta conformation and the pleated sheet).

Tertiary structure is conformational folding in addition to that imposed by secondary structure. Tertiary structure is stabilized by S—S bonds, ion-ion bonds, ion-dipole bonds, H bonds, and hydrophobic interactions between R groups. Hydrophobic R groups tend to cluster together in the center of most globular proteins.

Quaternary structure is the aggregation of protein subunits to form complexes. These complexes are stabilized by S—S bonds, ion-ion bonds, ion-dipole bonds, H bonds, and hydrophobic interactions between R groups.

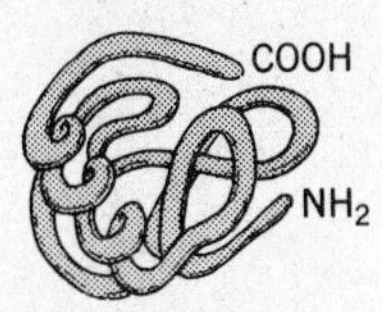

Globular protein

Aggregate functional protein

QUESTIONS AND PROBLEMS

SAMPLE RESPONSES

1) The diagram of insulin illustrates what is called the primary structure of proteins. The alpha helix illustrates secondary structure. Which (primary or secondary) structure is defined by listing the specific sequence of amino acids? Which involves the coiling of a single polypeptide chain, stabilized by H bonds?

–Primary
–Secondary
Note: Additional types of secondary structure, such as the pleated sheet and other helical patterns, do occur, but this text will restrict consideration to the α helix.

② How would you define secondary structure?

–Secondary sturcture involves the coiling of a polypeptide into an alpha helix that is stabilized by H bonding or involves other recurring patterns along one or two dimensions.

3) In most proteins only part of the polypeptide chain is coiled in the α-helix configuration.

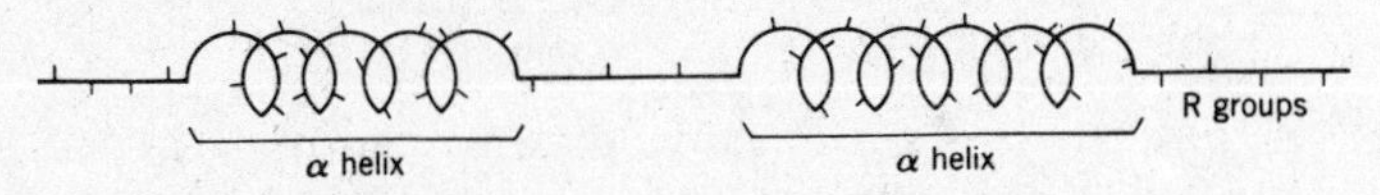

Label the two portions of the chain that exhibit secondary structure.

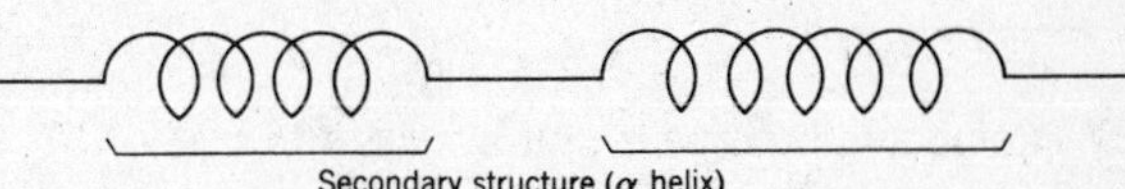

4) The preceding polypeptide chain can be folded and twisted further. (Compare with the preceding item)

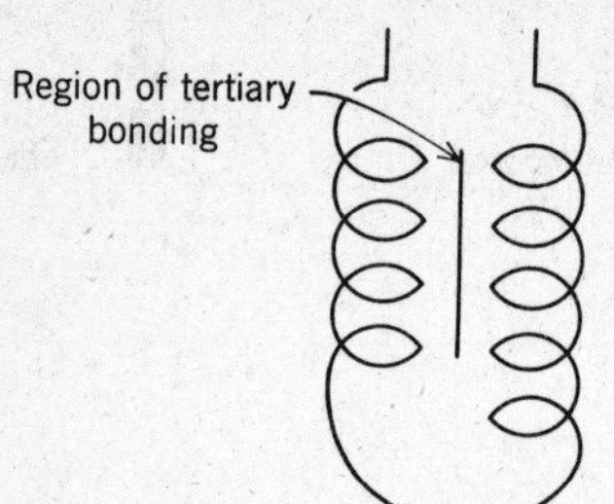

This is an example of *tertiary* structure. In tertiary bonding, are the amino acids involved close together (less than 5 amino acid units apart) or far apart (more than 5 amino acid units apart in the primary structure)?

–Far apart

5) Which amino acid can participate in the formation of disulfide (S–S) bonds?

–Cysteine

6) Which classes of amino acids have R groups that can participate in ion-ion bonding?

–Acidic (−) and basic (+) (plus the terminal $-COO^-$ and $-NH_3^+$ groups)

7) Which classes of amino acids have R groups that can act as dipoles in ion-dipole or H bonding?

–Hydroxy-polar, N-containing-polar, and S-containing-polar

8) In a polypeptide folded thus:

```
 /Phe-Gly-Leu-Val-Ile-Cys-Met-Asn- - - - -
His
 |
Ser
 \Gln-Val-Leu-Glu-Phe-Pro-Lys-Asp- - - - -
```

Which peptide units are most likely to have R groups that are oriented toward the "inside" of the folded chain? Explain your response.

–Isoleucine, valine, leucine, phenylalanine, valine, leucine, phenylalanine, and proline.
–These hydrocarbon R groups will cluster together toward the "inside" of the folded chain due to hydrophobic interactions. (or similar response)

(9) State a generalization about the orientation of hydrocarbon R groups in protein conformation.

–In proteins, hydrocarbon R groups are oriented toward the interior of the protein where they cluster together because of hydrophobic interactions.

10) List the types of bonding that stabilizes the tertiary structure of proteins.

–Disulfide (S–S), ion-ion, ion-dipole, H bonding (dipole-dipole), and hydrophobic interactions.

(11) Define tertiary structure in your own words.

–Tertiary conformation (structure) is the folding of a polypeptide chain in addition to that imposed by secondary structure. It is stabilized by S–S, ion-ion, ion-dipole, and H bonds and by hydrophobic interactions. (or similar response)

12) What types of bonding stabilize the quaternary structure of proteins?

–Disulfide, ion-ion, ion-dipole, H bonding, and hydrophobic interactions.

(13) Define quaternary structure.

–The quaternary structure of a protein refers to the aggregation of single chain subunits into a biologically functional complex. Quaternary structure is stabilized by S–S, ion-ion, ion-dipole, and H bands and by hydrophobic interactions. (or similar response)

CONCEPTS: CONJUGATED PROTEINS

Conjugated proteins consist of a polypeptide chain plus a nonpeptide group. For example, myoglobin (the oxygen storage pigment of red vertebrate muscle—the white meat of some fowl lacks myoglobin) is composed of globin (a polypeptide) plus heme (a nonpeptide iron containing group), as is diagrammed to the right.

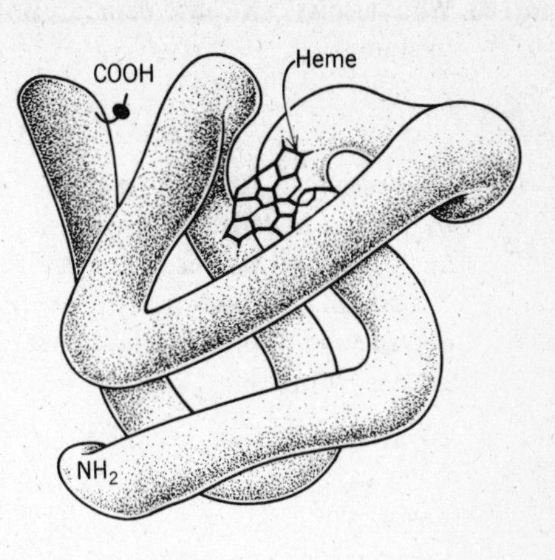

A *prosthetic group* is the nonpeptide group of a conjugated protein; for example, heme is the prosthetic group of myoglobin.

Denaturation of a protein involves the loss of biological activity because of changes in its tertiary or quaternary structure. Denaturation may be caused by heat, changes in pH, changes in salt concentration, heavy metal ions, and organic solvents (ether, alcohol, etc.)

A *basic protein* is positively charged at physiological pH because of a high percentage of lysine, arginine, or histidine.

Most proteins are acidic—negatively charged—at physiological pH.

QUESTIONS AND PROBLEMS

SAMPLE RESPONSES

14) What is the nature of the prosthetic group of the glycoproteins that were studied earlier?

—The carbohydrate chains (polysaccharides) that are attached to the protein are the prosthetic groups of glycoproteins.

15) Hemoglobin is the oxygen-carrying protein of red blood cells. Functional hemoglobin is an aggregate of four conjugated protein subunits; two subunits have identical alpha chains and two have identical beta chains. For hemoglobin, what does conjugated protein mean?

—Each protein subunit contains a nonprotein group.

What do you think is the nonprotein group of each subunit of hemoglobin?

—The iron containing heme group (as in myoglobin).

What is the general term for the heme groups of hemoglobin?

—Prosthetic groups

(16) Describe the factors that are involved in the primary, secondary, and tertiary, and quaternary structure of a hemoglobin molecule.

—Primary: the sequence of amino acids.
Secondary: the alpha helical portions of each polypeptide chain.
Tertiary: the specific conformational folding of each alpha and each beta chain (each with a heme group).
Quaternary: the aggregation of the 2 alpha and 2 beta chains to form the functional 4 unit hemoglobin molecule.

17) Egg albumin (egg white) is a protein that is shaped roughly like this in its natural state:

When an egg is cooked, the albumin proteins become shaped like this:

Reason out an explanation to account for the change from liquid to solid of egg white when it is heated.

—Cooking (heat) causes the denaturation of the albumin protein, changing the tertiary structure from a globular folded chain to an unfolded straight chain.

—R groups that were involved in intrachain bonding are now free to engage in bonding between chains.

—This chain-to-chain bonding produces the solid form of cooked egg white. (or similar response)

(18) What agents can cause denaturation of a functional protein?

–Heat, changes in pH, changes in salt concentrations, organic solvents (alcohol, ether), and heavy metal ions (mercury, lead, arsenic).

(19) Enzymes facilitate almost all the biochemical reactions that occur in cells. All enzymes are proteins. The normal functioning of each enzyme is dependent on a precise folding of the tertiary conformation of the polypeptide chain. Explain the effect of denaturation on enzyme function.

–Denaturation causes a change in tertiary conformation of the enzyme.
This change alters the normal functional structure of the enzyme, and biological activity is lost. (or similar response)

20) If a human has a high fever (over 108°F) for a few hours, irreversible cellular damage may occur. What is one possible explanation for such high temperature damage?

–High temperatures may cause protein (enzyme) denaturation, resulting in damage to cellular biochemical capabilities. (or similar response)

(21) Histones are a group of proteins that are associated with chromosomes. They have a high proportion of lysine, arginine, and histidine. What other general term would you apply to histones?

–Basic proteins

What charge would histones bear at physiological pH?

–Positive (+)

Take a 5-minute break before continuing.

SELF-TEST UNITS 11,12,&13

(Criterion score: 123 out of 129 total points)

(3 points) 1) Write the general structural formula for an amino acid.

(7 points) 2) List the 7 classes (groups) of amino acids.

(20 points) 3) Name the class to which each of the following amino acids is assigned:

Alanine	Leucine
Arginine	Lysine
Asparagine	Methionine
Aspartic acid	Phenylalanine
Cysteine	Proline
Glutamic acid	Serine
Glutamine	Threonine
Glycine	Tryptophan
Histidine	Tyrosine
Isoleucine	Valine

(5 points) 4) Name the 5 classes of amino acids whose R groups increase the water solubility of proteins.

(1 point) 5) Name the class of amino acids whose R groups decrease the water solubility of proteins.

(3 points) 6) Name the 3 classes of amino acids whose R groups contain polar groups.

(2 points) 7) Name the 2 classes of amino acids whose R groups are charged at physiological pH.

(1 point) 8) Name the class of amino acids whose R groups can participate in hydrophobic interactions.

(1 point) 9) Name the amino acid that participates in the formation of S–S bonds in proteins.

(6 points) 10) Identify each amino acid:

```
                                                                                 H
                                                                                 |
    H       H                                                                  H-C-H
    |       |                      H   H   H   H   H                           H  |  H   H
  H-C-------C-H                    |   |   |   |   |      O            O       |  |  |   |
    |       |   O              H2N-C---C---C---C---C---C               \\C----C--C--C---C-H
  H-C-H     C--C                   |   |   |   |   |    \              /      |  |  |   |
     \     / \  \                  H   H   H   H   NH2   OH          HO     H2N H  H   H
       N     H   OH
       |
       H

  O   OH
   \\ /           H                                       NH     H  H  H  H
    C             |                H  H      O             ||     |  |  |  |     O
    |        H    (benzene ring)   |  |     //         H2N-C---N--C--C--C--C--C
  H-C-NH2                      C---C--C--C                        |  |  |  |    \
    |                          ||     |   \                       H  H  H  H NH2  OH
  H-C-H      H                 C      H NH2 OH
    |             H        N       H
    SH                     |
                           H
```

(7 points) 11) Identify each amino acid:

(8 points) 12) Identify each amino acid:

(5 points) 13) Write the equation for the formation of a peptide bond between two amino acids of general structural formulae. Label the peptide bond.

(11 points) 14) In this structure:

Circle and identify by name each peptide unit.
What term identifies a peptide chain with this number of units?

(2 points) 15) What is the approximate molecular weight of insulin that contains 51 peptide units?

(4 points) 16) Describe the nature of the alpha helix.
Who first postulated this structure?

(3 points) 17) Explain the role of proline in relation to alpha helix formation in proteins.

(8 points) 18) In this polypeptide chain which peptide units might be charged + or − at physiological pH?

H_2N-ala-cys-val-his-asp-asn-his-phe-lys-arg-glu-phe-COOH

(5 points) 19) Explain electrophoresis in terms of the polypeptide in the preceding question.

(13 points) 20) Explain primary, secondary, tertiary, and quaternary structure in relation to hemoglobin. What types of bonds are important in stabilizing each level of conformation in proteins?

(2 points) 21) In general, what are the orientations of hydrocarbon R groups in a protein with folded conformation?

(2 points) 22) What is a prosthetic group?

(7 points) 23) What is denaturation?
What agents can produce denaturation?

(3 points) 24) Describe the nature of histones.

(129 points total)

SELF-TEST KEY UNITS 11,12,&13

(Criterion score 120 points: total of 126 points)

(3 points) 1)

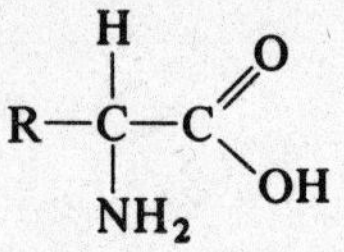

$$\begin{array}{ccccc} & & H & & O \\ & & | & & /\!/ \\ R & - & C & - & C \\ & & | & & \backslash \\ & & NH_2 & & OH \end{array}$$

(3 points–1 point off for each error) (*Review: Unit 11, Items 1-5*)

(7 points) 2) H-only, hydrocarbon, hydroxy-polar, S-containing-polar, N-containing-polar, acidic-ionized, and basic-ionized (1 point each) (*Review: Unit 11, Items 4-36*)

(20 points) 3)

ala: hydrocarbon	leu: hydrocarbon
arg: basic	lys: basic
asn: N-polar	met: S-polar
asp: acidic	phe: hydrocarbon
cys: S-polar	pro: hydrocarbon
glu: acidic	ser: OH-polar
gln: N-polar	thr: OH-polar
gly: H-only	trp: N-polar
his: basic	tyr: OH-polar
ile: hydrocarbon	val: hydrocarbon

(1 point each)

(*Review: Unit 11, Items 4-36, 46*)

(5 points) 4) OH-polar, N-polar, S-polar, acidic, and basic (one point each) (*Review: Unit 11, Items 4-36*)

(1 point) 5) Hydrocarbon (1) (*Review: Unit 11, Items 4-36*)

(3 points) 6) OH-polar, N-polar, and S-polar (1 point each) (Actually there are also dipolar groups in the acidic and basic amino acids) (*Review: Unit 11, Items 12-24*)

(2 points) 7) Acidic and basic (1 point each) (*Review: Unit 11, Items 25-34*)

(1 point) 8) Hydrocarbon (1) (*Review: Unit 11, Items 6-11*)

(1 point) 9) Cysteine (1) (*Review: Unit 11, Items 16-18*)

(6 points) 10) Proline (1), Lysine (1), Isoleucine (1), Cysteine (1), Tryptophan (1), Arginine (1) (*Review: Unit 11, Items 4-35*)

(7 points) 11) Aspartic acid (1), Glutamine (1), Histidine (1), Glycine (1), Cystine (1), Valine (1), Tyrosine (1) (*Review: Unit 11, Items 4-35*)

(8 points) 12) Serine (1), Leucine (1), Phenylalanine (1), Methionine (1), Asparagine (1), Glutamic acid (1), Threonine (1), Alanine (1) (*Review: Unit 11, Items 4-35*)

(5 points) 13)

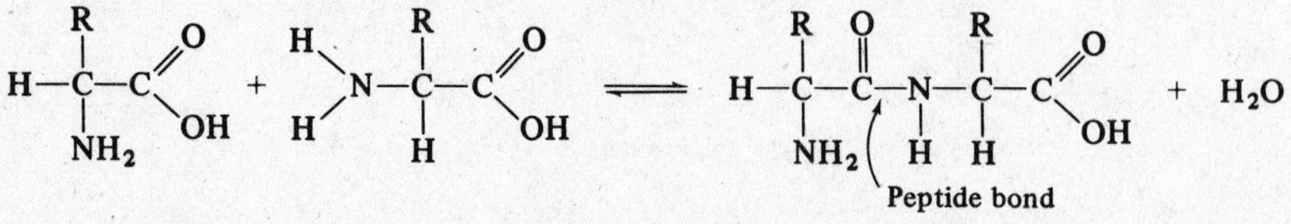

$$\begin{array}{ccccc} & & R & & O \\ & & | & & /\!/ \\ H & - & C & - & C \\ & & | & & \backslash \\ & & NH_2 & & OH \end{array} + \begin{array}{ccccccc} H & & & & R & & O \\ & \searrow & & & | & & /\!/ \\ & & N & - & C & - & C \\ & \nearrow & & & | & & \backslash \\ H & & & & H & & OH \end{array} \rightleftharpoons \begin{array}{ccccccccccc} & & R & & O & & & & R & & O \\ & & | & & \| & & & & | & & /\!/ \\ H & - & C & - & C & - & N & - & C & - & C \\ & & | & & & & | & & | & & \backslash \\ & & NH_2 & & & & H & & H & & OH \end{array} + H_2O$$

Peptide bond (the C—N bond between the C=O carbon and the N)

(5 points: 1 point off for each error) (*Review: Unit 12, Items 1-6*)

(11 points) 14)

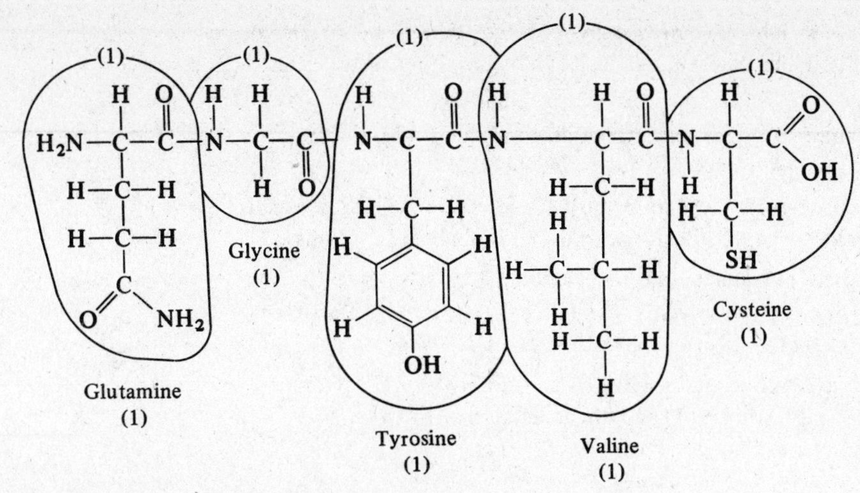

Pentapeptide (1) (*Review: Unit 12, Items 7-10*)

(2 points) 15) About 5100 daltons (2). (Each peptide unit contributes about 100 to the molecular weight.) (*Review: Unit 12, Item 11*)

(4 points) 16) An alpha helix is a helically coiled portion of a polypeptide chain (1) that is stabilized by H bonds (1). There are 3.6 peptide units per turn of the helix (1). Linus Pauling. (1) (*Review: Unit 12, Items 23-28*)

(3 points) 17) Proline terminates an alpha helix (1) because it cannot bend into the helical configuration (1) because of the participation of the amino group in the ring structure of the R group (1).
(*Review: Unit 12, Items 29-30*)

(8 points) 18)

ala terminal NH_2	+	
his	+	(1 point each)
asp	-	
his	+	
lys	+	
arg	+	
glu	-	
phe terminal COOH	-	(*Review: Unit 12, Items 18-19*)

(5 points) 19) Electrophoresis is the migration of molecules in an electric field. (2) It is due to the net charge of the polypeptide chain. (1) The above polypeptide has 5 + and 3 - charges; hence it will migrate toward the cathode. (2) (*Review: Unit 12, Items 18-22*)

(13 points) 20) Primary structure—the sequence of amino acids in polypeptide chains (1) that are genetically coded by DNA. (1)

Secondary structure—regions of alpha helix coiling in a polypeptide chain (1). Each polypeptide chain of hemoglobin has regions of alpha-helix coiling (1).

Tertiary structure—folding of the polypeptide in addition to that imposed by secondary structure. (1)

Quaternary structure—the aggregation of subunits (polypeptide chains) to form the functional protein. (1) Hemoglobin is composed of four conjugated protein subunits; each subunit contains a heme group.

Bonding: Primary—covalent (1)
Secondary—H bonding (1)
Tertiary and quaternary—disulfide (1), ion-ion (1), ion-dipole (1), H bonds (1), and hydrophobic interactions (1). (*Review: Units 12 and 13*)

(2 points) 21) Hydrocarbon R groups are usually oriented toward the interior of globular proteins where they cluster in hydrophobic interactions. (2) (*Review: Unit 13, Items 8-9*)

(2 points) 22) A prosthetic group is the nonpeptide component (1) of a conjugated protein. (1) (*Review: Unit 13, Items 14-15*)

(7 points) 23) Denaturation is a change in tertiary conformation of a functional protein that results in the loss of normal biological functions. (2)

Agents: heat (1), changes in pH (1), changes in salt concentrations (1), organic solvents (alcohol, ether, etc.) (1), heavy metal ions (1). (*Review: Unit 13, Items 17-20*)

(3 points) 24) Histones are basic proteins (1) that contain high levels of lysine, arginine, and histidine (1) and are associated with DNA in chromosome structure (1). (*Review: Unit 13, Item 21*)

Take at least a 5-minute break before continuing.

UNIT FOURTEEN — LIPIDS: FATS, PHOSPHOLIPIDS, STEROLS — 14

CONCEPTS

Fats are important physiologically as food storage molecules and as thermal (heat) insulation in warm-blooded animals. The related phospholipids are essential components of biological membranes. Another group of fatlike molecules, the sterols, function as membrane constituents, as hormones, and in fat digestion.

Lipid: a general term for all fatlike substances. Lipids are soluble in ether, chloroform, and other organic solvents.

Fatty acids are composed of a hydrocarbon chain plus a carboxyl group.

Structure of a fatty acid

$$H-(CH_2)_{16}-C(=O)OH \qquad \text{(Hydrocarbon)}-C(=O)OH$$

Synthesis and Breakdown of a Fat

Glycerol + 3 fatty acids ⇌ Fat molecule + $3\ H_2O$ (3 water)

General Structure of a Phospholipid

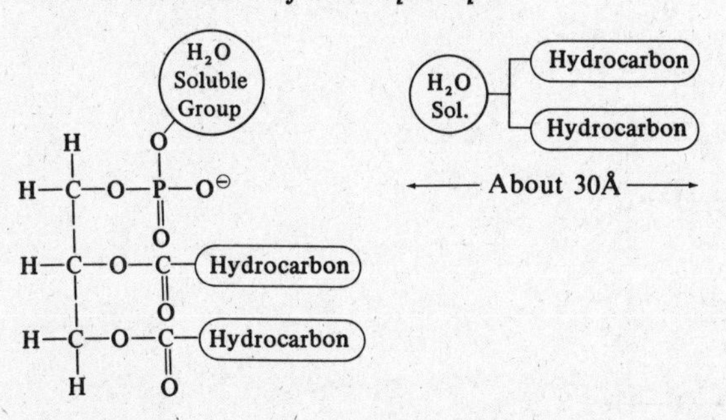

Lecithin: A Typical Phospholipid

In lecithin the water soluble group is choline. In other phospholipids the water soluble groups include serine, ethanolamine, and inositol (a sugar).

Do not attempt to memorize this material before going on to the questions and problems section.

QUESTIONS AND PROBLEMS

SAMPLE RESPONSES

1) What two groups participate in an ester linkage?

–Carboxyl $\left(-C\begin{smallmatrix}=O\\OH\end{smallmatrix}\right)$ (organic acid) (any order)

Hydroxyl (–OH)

2) Circle the hydroxyl groups.

```
    H                    H
    |                    |
 H—C—OH             —H—C—(OH)
    |                    |
 H—C—OH              H—C—(OH)
    |                    |
 H—C—OH              H—C—(OH)
    |                    |
    H                    H
```

3) Fatty Acids

```
   H  H                  H  H  H
   |  |     O            |  |  |     O
H—C—C—C//             H—C—C—C—C//
   |  |   \              |  |  |   \
   H  H    OH            H  H  H    OH

   H  H  H  H  H  H  H
   |  |  |  |  |  |  |     O
H—C—C—C—C—C—C—C—C//
   |  |  |  |  |  |  |   \
   H  H  H  H  H  H  H    OH
```

Of what two groups is a fatty acid composed?

–Hydrocarbon
Carboxyl (organic acid) (any order)

4) If the general formula for a fatty acid is: $R—C(=O)OH$

what does R equal?

–R = a hydrocarbon group

5) Form an ester linkage between the *top* C of glycerol and the fatty acid

```
    H                                 H    O
    |                                 |    ||
 H—C—OH           O              —H—C—O—C—R
    |        R—C//                    |
 H—C—OH          \                H—C—OH
    |             OH                  |
 H—C—OH                           H—C—OH
    |                                 |
    H                                 H
```

6) Similarly, form ester linkages with the remaining two C's of glycerol and two more molecules of fatty acid.

```
    H    O
    |    ||
 —H—C—O—C—R
    |    O
    |    ||
  H—C—O—C—R
    |    O
    |    ||
  H—C—O—C—R
    |
    H
```

7) You have just constructed a *fat* (or biological oil). In the "R" of fats, C = 15 or 17 (usually). Write out the *complete* structural formula for a fat made up of stearic acid ($C_{17}H_{35}COOH$). (All fats contain glycerol.)

```
    H    O  H  H  H  H  H  H  H  H  H  H  H  H  H  H  H  H  H
    |    || |  |  |  |  |  |  |  |  |  |  |  |  |  |  |  |  |
 —H—C—O—C—C—C—C—C—C—C—C—C—C—C—C—C—C—C—C—C—C—H
    |       |  |  |  |  |  |  |  |  |  |  |  |  |  |  |  |  |
    |       H  H  H  H  H  H  H  H  H  H  H  H  H  H  H  H  H
    |    O  H  H  H  H  H  H  H  H  H  H  H  H  H  H  H  H  H
    |    || |  |  |  |  |  |  |  |  |  |  |  |  |  |  |  |  |
  H—C—O—C—C—C—C—C—C—C—C—C—C—C—C—C—C—C—C—C—C—H
    |       |  |  |  |  |  |  |  |  |  |  |  |  |  |  |  |  |
    |       H  H  H  H  H  H  H  H  H  H  H  H  H  H  H  H  H
    |    O  H  H  H  H  H  H  H  H  H  H  H  H  H  H  H  H  H
    |    || |  |  |  |  |  |  |  |  |  |  |  |  |  |  |  |  |
  H—C—O—C—C—C—C—C—C—C—C—C—C—C—C—C—C—C—C—C—C—H
    |       |  |  |  |  |  |  |  |  |  |  |  |  |  |  |  |  |
    H       H  H  H  H  H  H  H  H  H  H  H  H  H  H  H  H  H
```

8) Look carefully at the fat you have just diagrammed. Would you expect this fat to be soluble in water? Explain you answer.

–No. The hydrophobic groups (nonpolar or hydrocarbon groups) promote insolubility. (or similar answer)

(9) Which of the following would be the *least* soluble in H_2O?

A. $C_{21}H_{43}COOH$
B. $C_{17}H_{35}COOH$
C. $C_5H_{11}COOH$

–A

(10) Would several fat molecules cluster together in water? Explain your answer.

–Yes. The hydrocarbon groups would cluster together in hydrophobic interactions.

(11) Diagram the structure of a fat composed of:

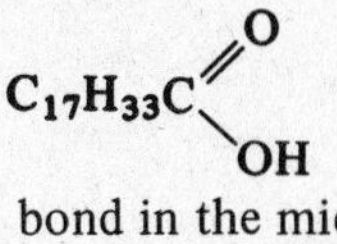

$C_{17}H_{33}C(=O)OH$ (oleic acid) using the empirical formula. Oleic acid contains a double bond in the middle of the carbon chain, for example:

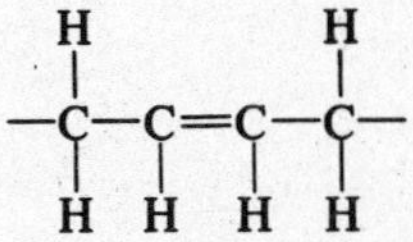

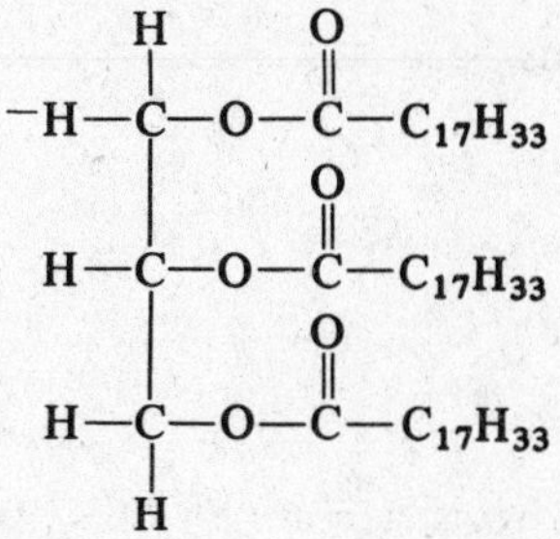

$$-\underset{H}{\overset{H}{C}}-\underset{H}{C}=\underset{H}{C}-\underset{H}{\overset{H}{C}}-$$

$$\begin{array}{l} H-\overset{H}{C}-O-\overset{O}{\overset{\|}{C}}-C_{17}H_{33} \\ H-C-O-\overset{O}{\overset{\|}{C}}-C_{17}H_{33} \\ H-\underset{H}{C}-O-\overset{O}{\overset{\|}{C}}-C_{17}H_{33} \end{array}$$

12)

$$\begin{array}{l} H-\overset{H}{C}-O-\overset{O}{\overset{\|}{C}}-C_{17}H_{35} \\ H-C-O-\overset{O}{\overset{\|}{C}}-C_{17}H_{35} \\ H-\underset{H}{C}-O-\underset{O^{\ominus}}{\overset{O}{\overset{\|}{P}}}-O-R_s \end{array}$$

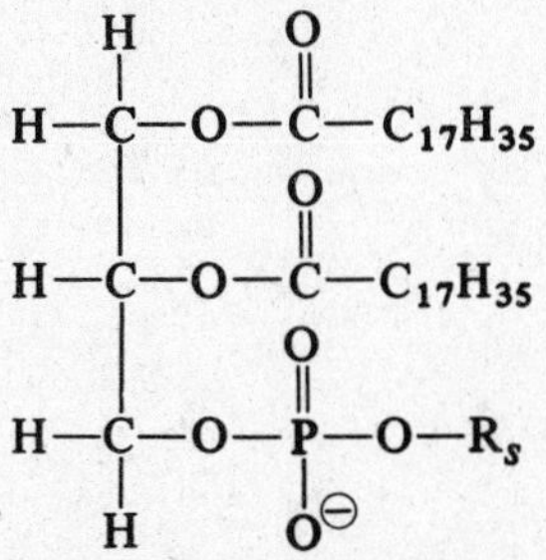

This is the structural formula of a typical *phospholipid*. What two groups of the phospholipid replace one of the fatty acids in its related fat?

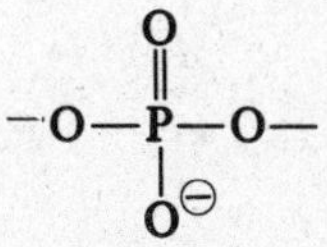

$-O-\underset{O^{\ominus}}{\overset{O}{\overset{\|}{P}}}-O-$ (phosphate)

R_s (R_s = a water soluble group)

13) In the phospholipid lecithin the R_s group is choline:

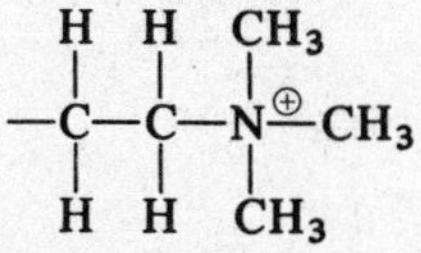

$$-\underset{H}{\overset{H}{C}}-\underset{H}{\overset{H}{C}}-\underset{CH_3}{\overset{CH_3}{N^{\oplus}}}-CH_3$$

Is an ion part of this R_s group?

–Yes (+ charge)

Would this R_s group promote solubility in water? Explain you answer.

–Yes. It would attract water molecules because of ion-dipole binding.

14) On the diagram of the phospholipid (item 12) circle and label the portions of the molecule that are *hydrophobic* (water hating). Also circle and label the portions of the molecule that are *hydrophilic* (water loving).

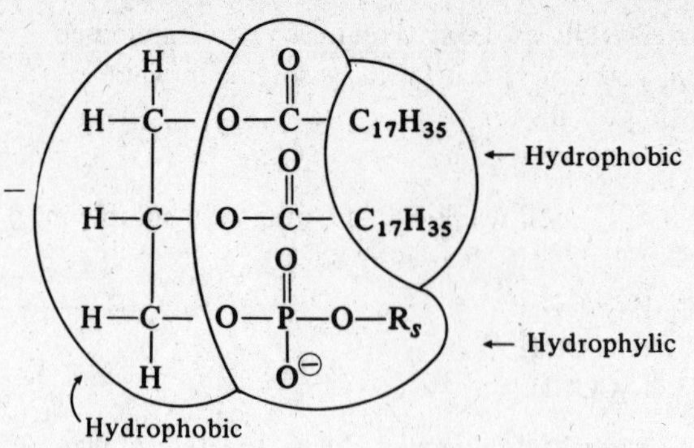

— H—C—O—C—$C_{17}H_{35}$ ← Hydrophobic
H—C—O—C—$C_{17}H_{35}$
H—C—O—P—O—R_s ← Hydrophylic
Hydrophobic

(15) Diagram the general structure of a phospholipid. Use

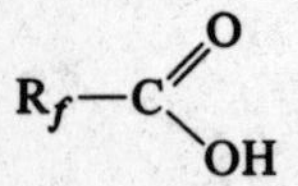

R_f—C(=O)OH = the fatty acids

R_s = the H_2O soluble group

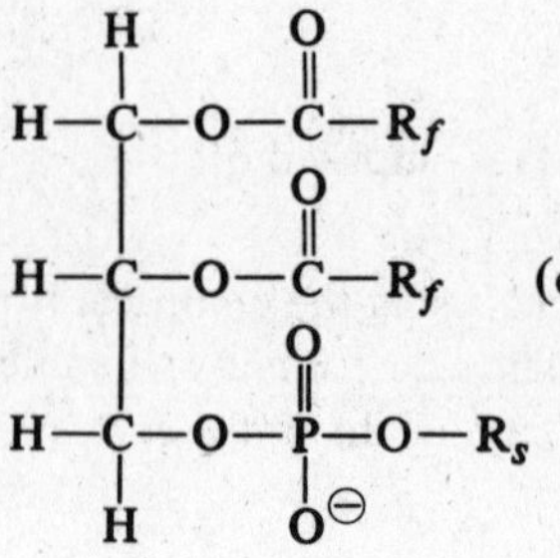

— H—C—O—C(=O)—R_f
H—C—O—C(=O)—R_f (or similar structures)
H—C—O—P(=O)(O⊖)—O—R_s

(16) Phospholipids are frequently written in ultrashorthand form thus:

Label the hydrophobic groups and the hydrophilic group on the shorthand diagram.

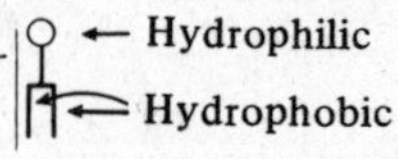

— ← Hydrophilic
← Hydrophobic

17) Diagram five (shorthand form) phospholipid molecules to indicate the clustering together of the hydrophobic groups in water.

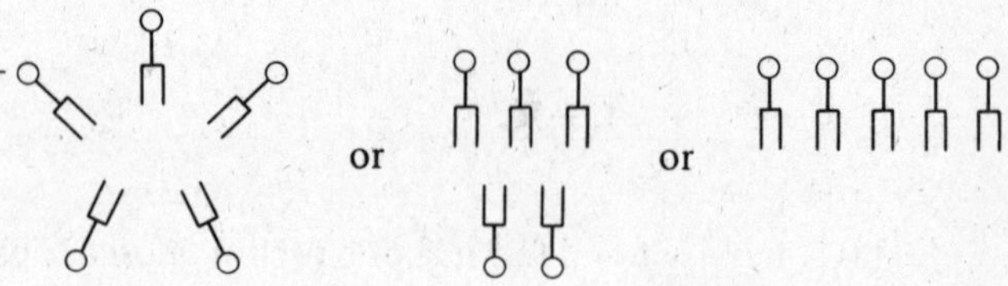

— or or

(or similar configuration)

CONCEPTS: STEROLS

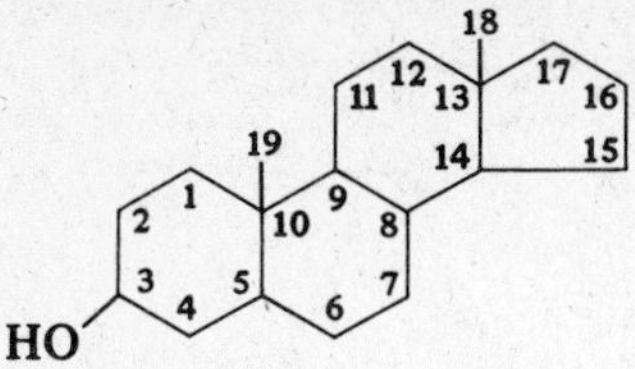

Sterol nucleus

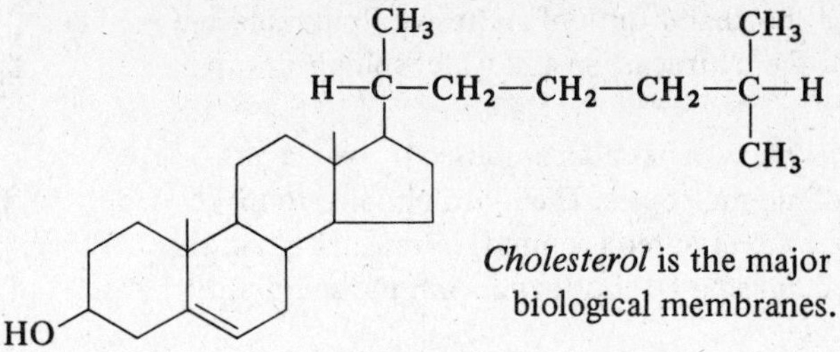

Cholesterol is the major sterol of biological membranes.

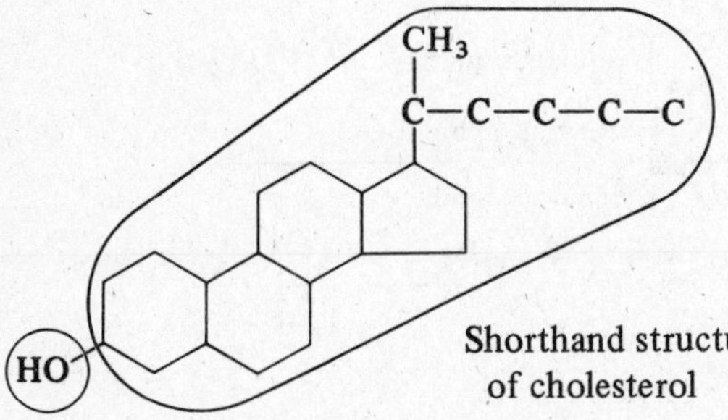

Shorthand structure of cholesterol

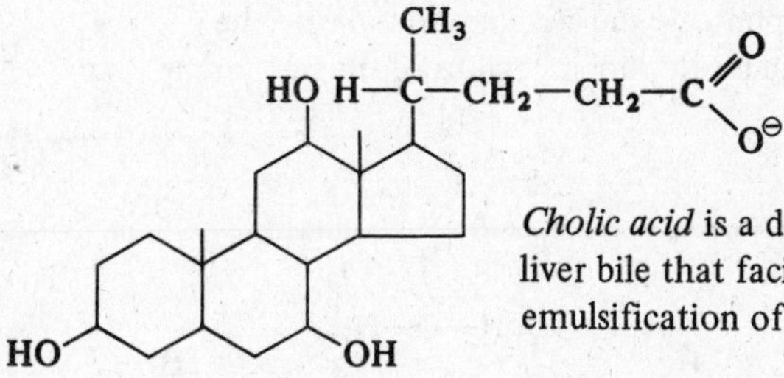

Cholic acid is a detergent in liver bile that facilitates the emulsification of fats in digestion.

Testosterone is an *androgen* or male steroid hormone. It is produced by the testes and adrenal glands and controls the development of secondary sex characteristics, such as beard and hair growth, muscle development, stature, voice tone, and male behavior.

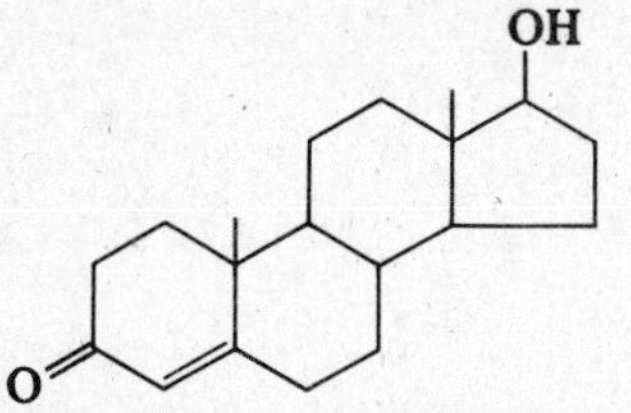

Aldosterone is produced by the cortex of the adrenal gland and functions in regulating salt balance. It promotes the retention of salts by the kidney, so that less salt is excreted in the urine.

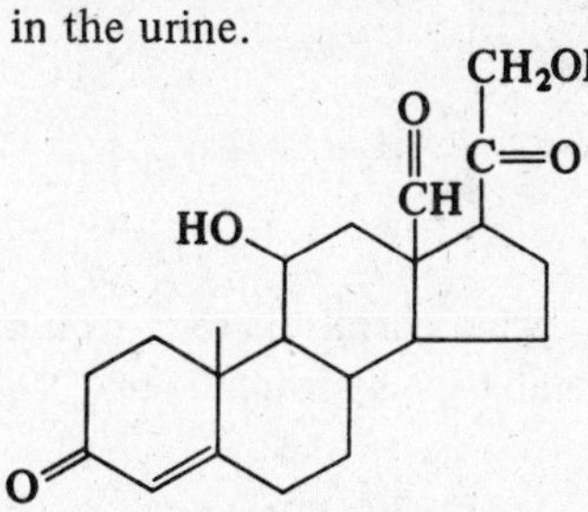

Estrone is an *estrogen* and is one of the female steroid hormones produced by follicle cells of the ovary and by the adrenal glands. It controls the development of mammary glands, subcutaneous fat deposition, pelvic configuration, uterus development, and female behavior patterns.

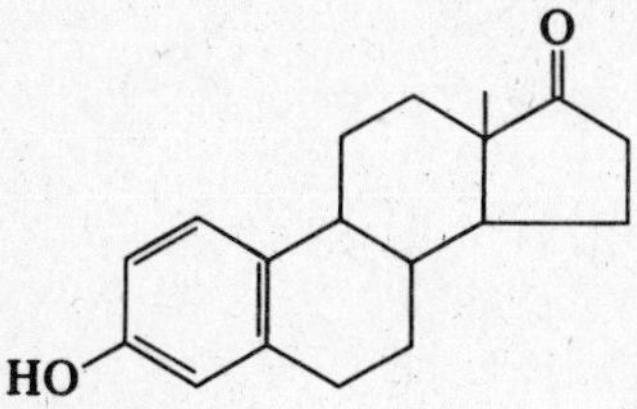

Corticosterone is also produced by the adrenal cortex. It is a *glucocorticosteroid*, a group of hormones that regulate carbohydrate metabolism in all cells.

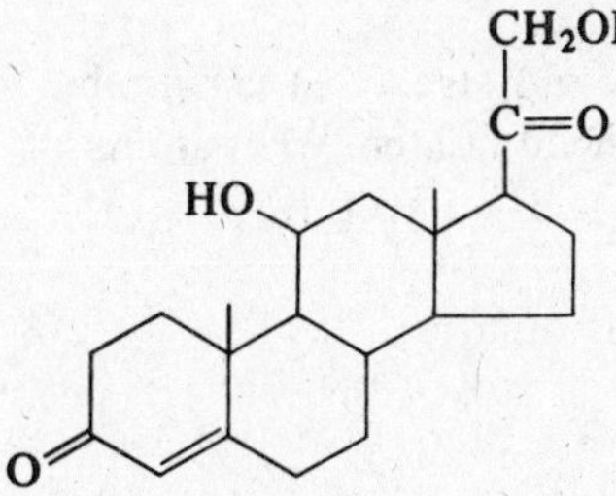

Progesterone is produced by the corpus luteum of the ovary and by the adrenal cortex. It regulates uterine development, pregnancy, and the onset of labor.

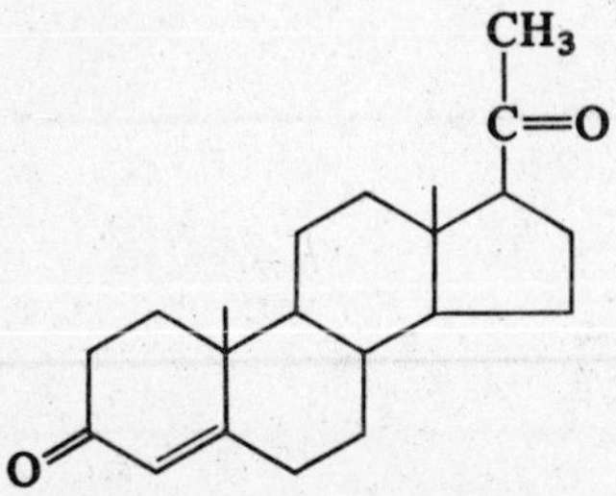

The *prostaglandins* are a class of closely related molecules that function in short range intercellular communication and in intracellular communication. They influence smooth muscle contractions, especially of blood vessels, and also affect the permeability of cell membranes to water and salts.

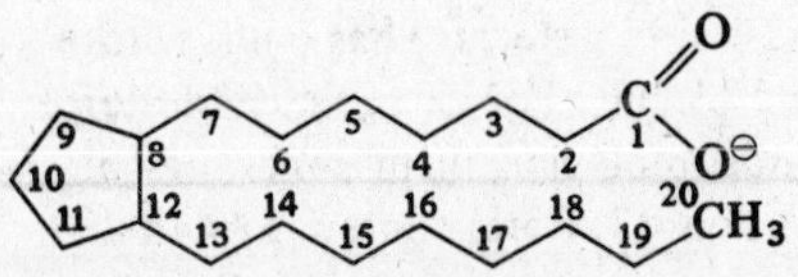

QUESTIONS AND PROBLEMS

SAMPLE RESPONSES

18) Write the shorthand form of a phospholipid molecule and label the hydrocarbon and water-soluble groups.

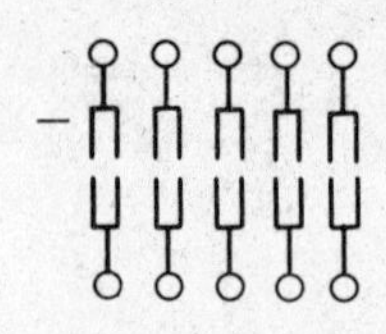

19) Biological membranes are composed of two layers of phospholipid molecules. The hydrophobic groups of each layer are oriented (pointed) inward. Diagram this relationship using 10 shorthand form phospholipid molecules.

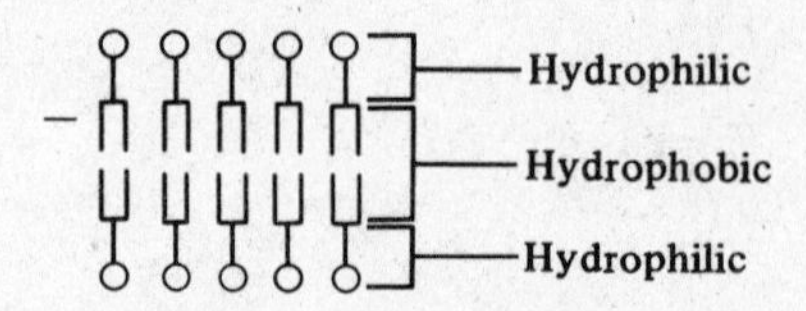

This is the basic structure of all biological membranes.

20) Label your response to the last item to indicate the hydrophilic and hydrophobic regions of the membrane lipid bilayer.

Hydrophilic
Hydrophobic
Hydrophilic

21) Steroid nucleus:

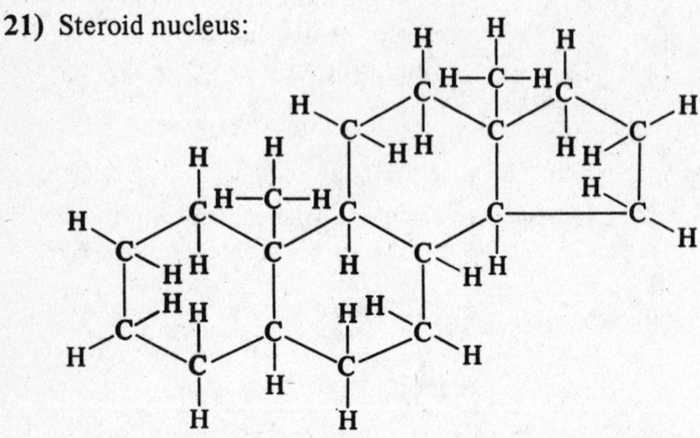

Is this steroid aromatic? In what chemical group would you place this steroid? Would it be hydrophilic or hydrophobic?

–No
–Hydrocarbon (cyclic)
–Hydrophobic

22)

The steroid nucleus (basic ring structure) can also be diagrammed in this shorthand fashion. What atom is located at each intersection or point? (arrows at A)

–C (carbon)

23) Each C atom is assigned a number.

How many H's are bonded to carbon #1? 5? 17? 4? 8? 15?

–#1. 2H 5. 1H 17. 2H
4. 2H 8. 1H 15. 2H

24) A ster*ol* (as contrasted to a ster*oid*) has one or more OH groups in the molecule. For example, cholesterol, which occurs in most biological membranes, has an –OH group on carbon #3. Refer to item 23 and diagram a shorthand steroid nucleus with this OH added.

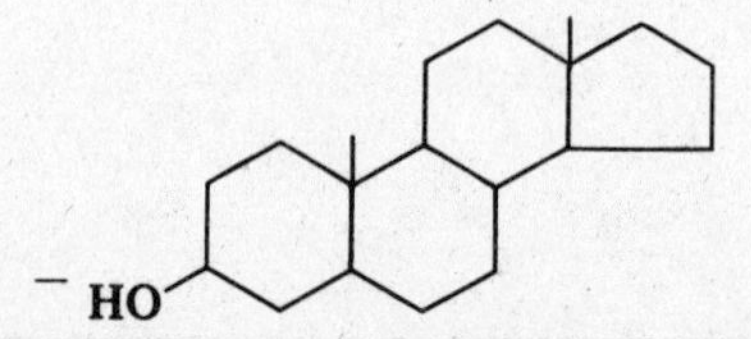

25) In cholesterol, there are also other differences from the basic steroid nucleus. Diagram these on your outline diagram (item 24). There is a double bond between carbons 5 and 6. A hydrocarbon chain

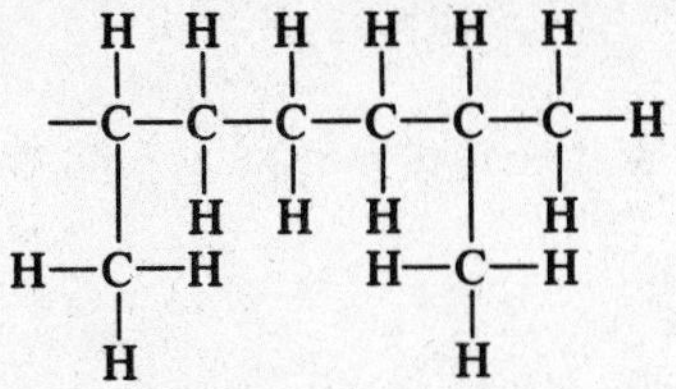

```
 H  H  H  H  H  H
 |  |  |  |  |  |
-C--C--C--C--C--C-H
 |  |  |  |  |  |
 |  H  H  H  |  H
H-C-H     H-C-H
 |           |
 H           H
```

is attached to carbon #17

–

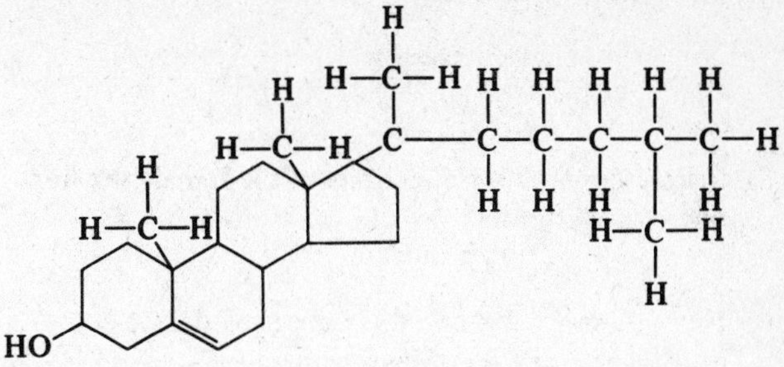

```
                   H
                   |
              H  H-C-H H  H  H  H  H
              |    |   |  |  |  |  |
            H-C-H  C---C--C--C--C--C-H
     H                 |  |  |  |  |
     |                 H  H  H  |  H
   H-C-H                      H-C-H
                                 |
                                 H
HO
```

26) Which portion of the cholesterol molecule would be *hydrophilic*? Sterols are a common constituent of biological membranes. Would you expect to find them in the lipid or protein layer?

– –OH

–Lipid (due to hydrophobic bonding)

27) The lipid layer of most biomembranes contains roughly 80% phospholipids and 20% sterols. Using this shorthand diagram for cholesterol, diagram the relationship between 8 phospholipid molecules and 2 sterol molecules in a lipid bilayer biomembrane.

–

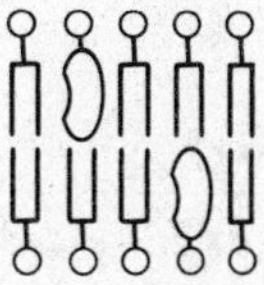

(28) In your own words, describe the basic structure of biological membranes.

–Biomembranes are composed of a lipid bilayer. The composition is roughly 80% phospholipids and 20% sterols.

(29) Diagram the basic ring structure of a steroid.

–

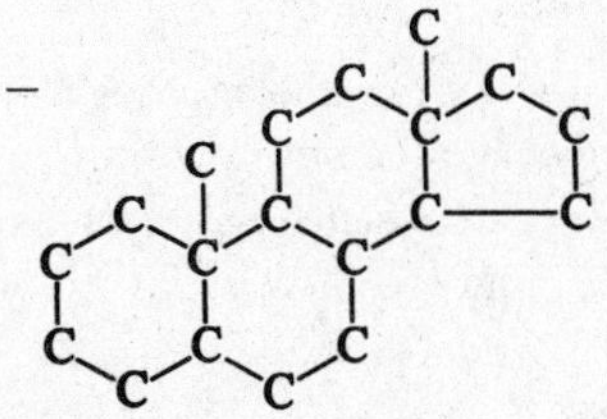

or

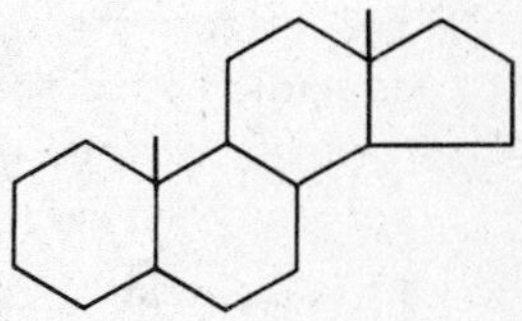

30) Examine the structure of cholic acid in the Concepts page. Explain how it might function to promote the solubility of fats in water.

–Cholic acid contains both hydrophobic and hydrophilic regions (3 OH + $COO^{\ominus}$). Hence it will attract fat molecules by hydrophobic interactions and will itself be soluble in water because of its hydrophilic groups.

(31) What is the function of cholic acid?

–Cholic acid is a constituent of liver bile; it functions as a detergent in the emulsification of fats in digestion.

32) Examine the structural formulae for the three steroid sex hormones in the Concepts section.
Are all three sterols?
In what general ways do the three molecules differ?

–No, only estrone has an OH group on C #3.
–OH vs. =O on C #3; the number and position of double bonds in the first ring; the group attached to C #17.

(33) Briefly describe the functions of the male steroid sex hormones, the androgens.

–Control the expression of the male secondary sex characteristics: beard and hair growth, muscle development, stature, voice tone, and male behavior.

(34) Briefly describe the functions of the female sex hormones, the estrogens.

–Control the expression of the female secondary sex characteristics: development of mammary glands, subcutaneous fat deposition, pelvic configuration, uterine development, and female behavior.

(35) Briefly describe the functions of the female sex hormone progesterone.

–Regulation of uterine development, pregnancy, and the onset of labor in childbirth.

(36) From memory, list the three types of steroid sex hormones that function in mammals.

–Androgens, estrogens, and progesterone.

37) Examine the structural formulae of the two adrenal cortex steroid hormones that are illustrated in Concepts. What groups do they have in common that are not contained by any of the sex steroids?

–An OH on C #11.
An OH on the terminal C of the group that is attached to C #17.

38) What is the major function of aldosterone?

–It promotes the retention of salts by the kidney (hence, it promotes salt retention by the blood and tissue fluid of the body).

39) What is the major function of the glucocorticosteroids?

–They regulate carbohydrate metabolism in all cells.

(40) Name two adrenal cortical steroid hormones and give the major function of each.

–Aldosterone: regulates salt balance.
Glucocorticosteroids (corticosterone): regulates carbohydrate metabolism in all cells.

41) Which would be least soluble in water: cholesterol, the sex steroids, or the adrenal steroids?

–Cholesterol (which has only one polar group).

42) The various prostaglandins differ from one another primarily by the positioning of double bonds within the molecules and the presence or absence of OH or =O groups on carbons, 9, 11, and 15.
What group involves C #1?
What group involves C #20?
Are the prostaglandins steroids?

–Carboxyl
–Methyl
–No

(43) What are the major functions of the prostaglandins?

–Short range intercellular communication
–Intracellular communication
–Regulation of smooth muscle contraction, particularly of blood vessels
–Control permeability of cell membranes to water and salts

(44) List four types of steroids that you have studied.

–Cholesterol: membrane steroid
Cholic acid: a bile steroid
Sex hormones
Adrenal corticosteroids

Take at least a 5-minute break before continuing to the next unit.

UNIT FIFTEEN NUCLEIC ACIDS 15

CONCEPTS

The nucleic acids function in living organisms as genetic material and in the control of protein (hence enzyme) synthesis. Related polyphosphate molecules (ATP, GTP, CTP) function in energy transfer within organisms. This unit will attempt only to introduce you to the rudiments of nucleic acid structure. The roles of DNA as genetic material and RNA as the machinery of protein synthesis are covered in the companion volume *Concepts in Molecular Genetics and Development*.

Nucleotides are the structural units of nucleic acids. Each nucleotide is composed of three subunits:

Adenosine monophosphate

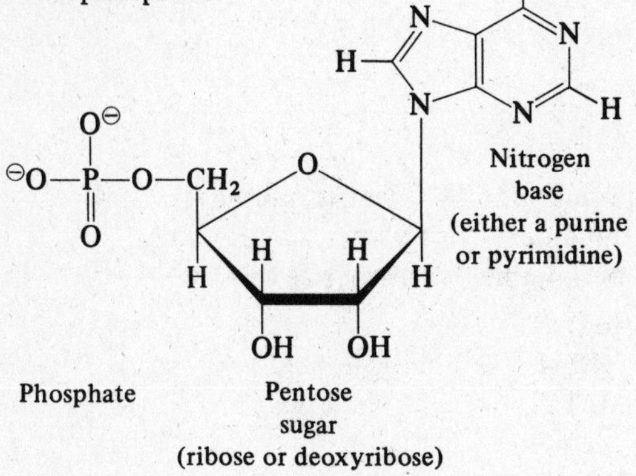

Phosphate

Pentose sugar (ribose or deoxyribose)

Nitrogen base (either a purine or pyrimidine)

Nucleotides are usually identified by their N-bases (A, G, U, T, C):

Purines have 2 rings; common purines:

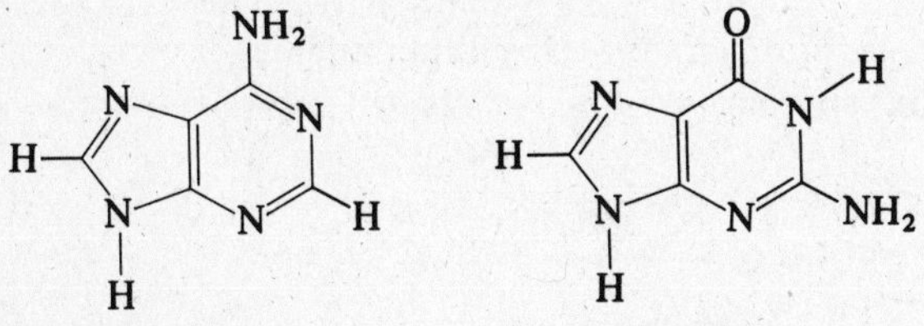

Adenine

Guanine

Pyrimidines have 1 ring; common pyrimidines:

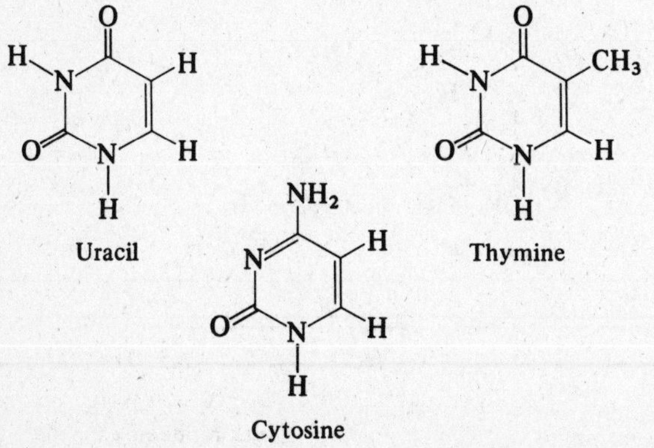

Uracil

Thymine

Cytosine

Nucleotides can be joined by phosphate ester linkages to form nucleic acids:

Single stranded DNA

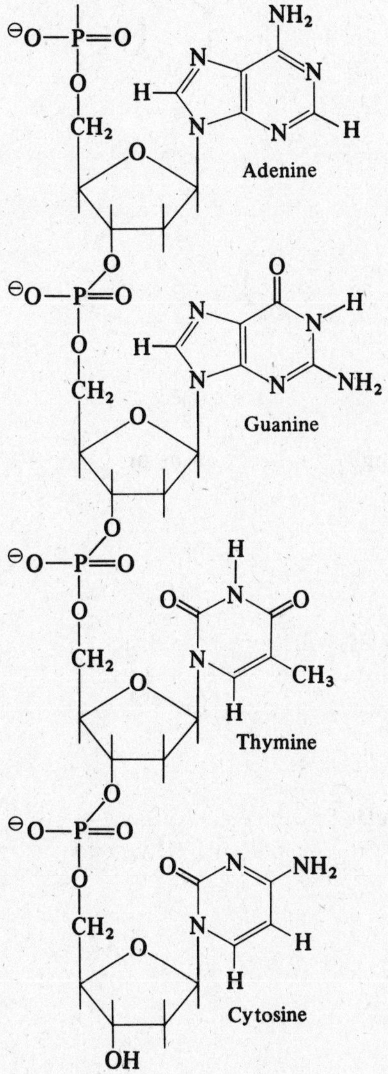

Adenine

Guanine

Thymine

Cytosine

Shorthand structures

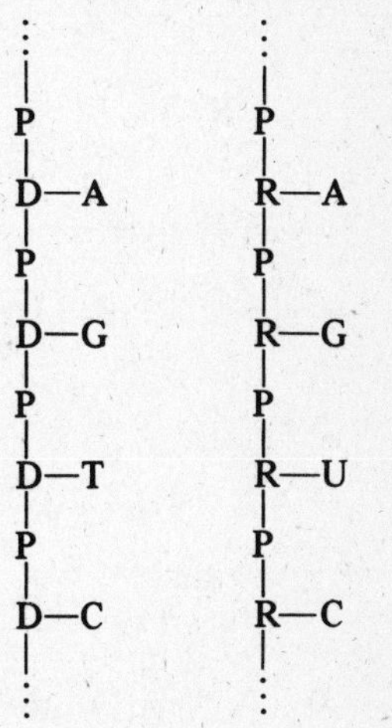

for DNA	for RNA
⋮	⋮
P	P
D—A	R—A
P	P
D—G	R—G
P	P
D—T	R—U
P	P
D—C	R—C
⋮	⋮

QUESTIONS AND PROBLEMS

SAMPLE RESPONSES

1)

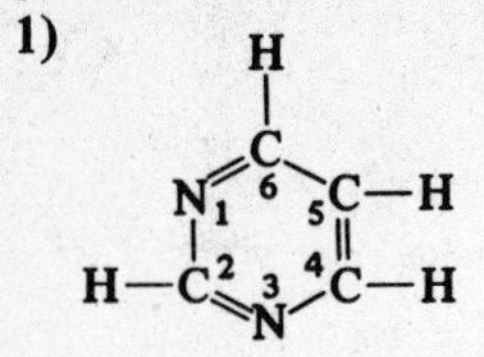

Pyrimidine ring

Purine ring

Write a shorthand diagram for the 2 nitrogen bases, the purine and pyrimidine rings. (Remember that *only* C and H are *not* written in a shorthand diagram.)

Pyrimidine (1 ring)

Purine (2 rings)
(Pu-rymes with 2)

–Pronunciation:
PIER-IM′-U-DEAN (IM- as in him; U as in up)
PURE′-EEN

2) Without looking back to item 1, label each of the following N-bases as a purine or pyrimidine.

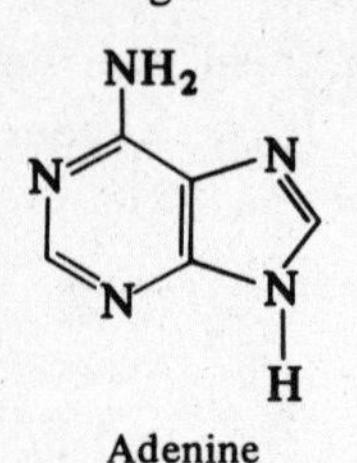

Adenine

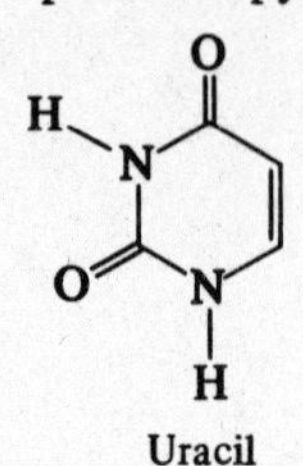

Uracil

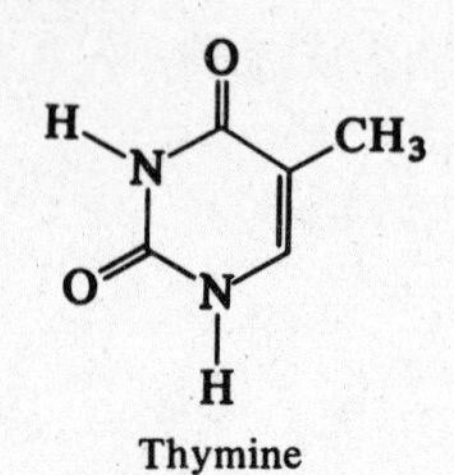

Thymine

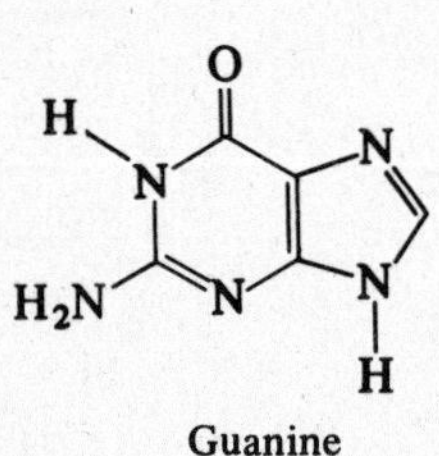

Guanine

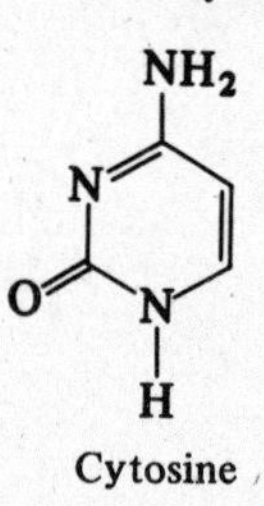

Cytosine

–Adenine: purine
Uracil: pyrimidine
Thymine: pyrimidine
Guanine: purine
Cytosine: pyrimidine
(memorize this list)

–Pronunciation:
ADD′-U-NEEN (U as in up)
EUR′-U-SILL
THIGH′-MEAN
GUA′-NEEN (GUA as in guano)
CY′-TOE-SEEN

3) Label each of the following N-bases as a purine or pyrimidine.

Adenine Thymine
Cytosine Uracil
Guanine

–Adenine: purine
Cytosine: pyrimidine
Guanine: purine
Thymine: pyrimidine
Uracil: pyrimidine

4) In the unit on carbohydrates, you learned the structural formula for two pentoses. What were these two pentoses? (If you cannot recall, see items 1 and 2, Unit 10.)

–Deoxyribose
–Ribose
} any order

5) Write the structural formulae for ribose and deoxyribose. Number the C atoms of one molecule. (If you do not remember, guess.)

Ribose

Deoxyribose
(Check numbers carefully)

6) Write the structural formula for ribose in the mirror image form (C #1 on the left and C #5 on the right).

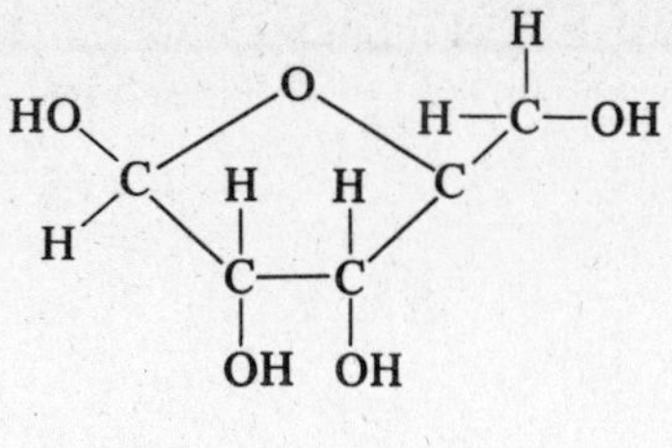

–[structural formula of ribose, mirror image form]

7) Attach a molecule of ribose (in the mirror image form) to this molecule of adenine by a covalent bond between atom #9 of the adenine (see item 1) and C #1 of the ribose. What molecule is "split out" in the formation of this covalent bond?

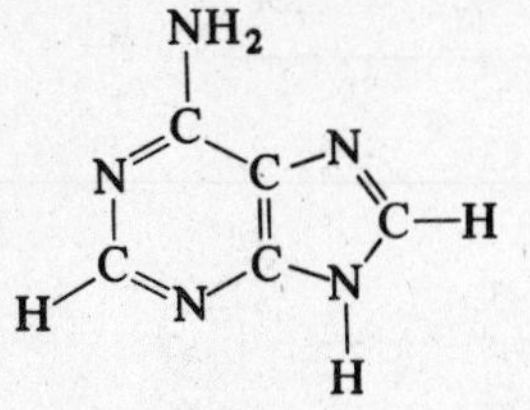

Adenine

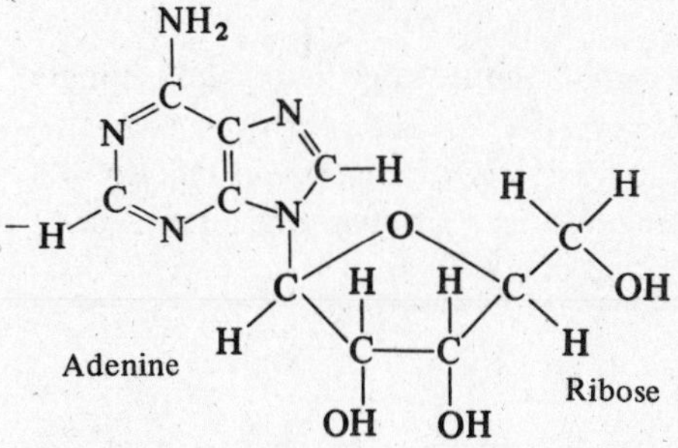

–[structural formula: Adenine attached to Ribose]

Adenine

Ribose

–Water is split out.

8) To the molecule you diagrammed in the previous item, add a phosphate group via an *ester linkage* to C #5 of the ribose component. (This type of ester linkage will be referred to as a phosphate ester linkage since it involves phosphorous rather than a carbon atom.)

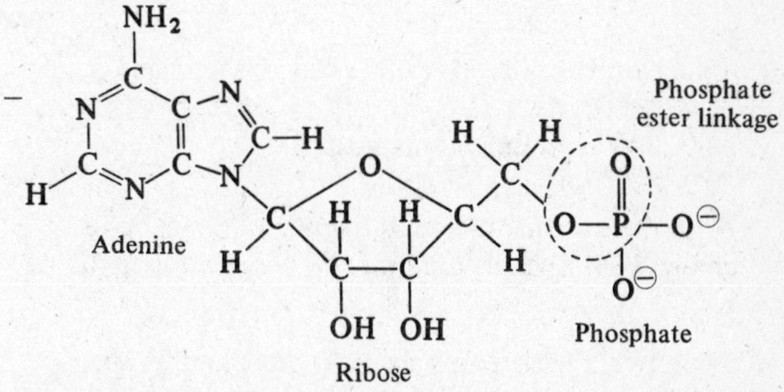

–[structural formula: Adenine–Ribose–Phosphate]

Adenine

Phosphate ester linkage

Phosphate

Ribose

(This molecule is a *nucleotide, adenosine monophosphate.*)

What molecule is "split out" in the formation of a phosphate ester linkage?

–A water molecule is split out. (The second H comes from H^+ in solution.)

9) In similar fashion the other N-bases, cytosine, guanine, thymine, and uracil will form nucleotides. Also, the other pentose sugar, deoxyribose, can form nucleotides. In this nucleotide, circle and label the three basic components.

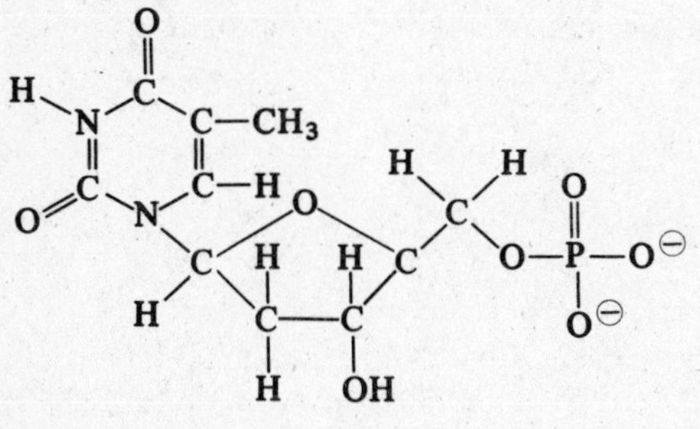

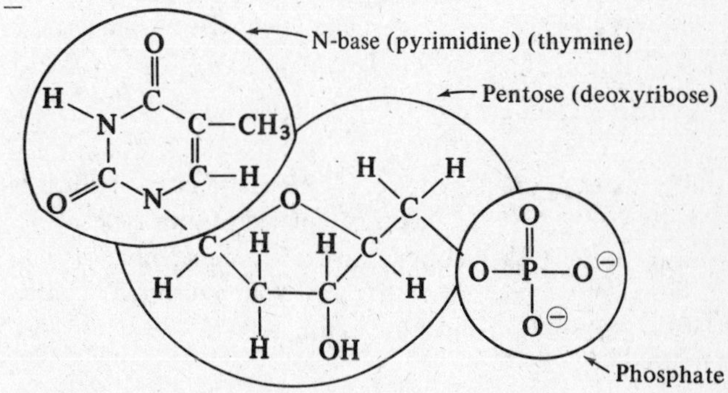

[structural formula of thymine deoxyribose nucleotide]

–[same structure with components circled]

N-base (pyrimidine) (thymine)

Pentose (deoxyribose)

Phosphate

(10) What are the three components of which a nucleotide is composed?

–N-base (purine or pyrimidine)
Pentose sugar (ribose or deoxyribose)
Phosphate
(any order)

11) Diagram the *shorthand* structural formula for the nucleotide, adenosine monophosphate. Use the shorthand diagram for adenine =:

NH₂ — adenine ring (N, N, N, N–H)

NH₂ — adenine ring linked through N to ribose (O; OH OH) —O—P(=O)(O⊖)—O⊖

12) Two nucleotides can be linked together by a phosphate ester linkage between the P of one nucleotide and sugar C #3 of the other nucleotide. Diagram this linkage. Diagram only the phosphate group of one nucleotide and the deoxyribose of the other.

—(N-base)—C(H) ring with O, C(H)(H), C(H)—O—P(=O)(O⊖)—O—(pentose)—(N-base); CH₂—O—P(=O)(O⊖)—O⊖

(or other reasonable forms of the same ester linkage)

13) A nucleotide can be written in (ultra) shorthand form:

P—R
  | = Adenosine monophosphate
  A

Write the shorthand form for thymidine monophosphate. What does R stand for? What would D stand for?

– P—R
      |
      T

–Ribose

–Deoxyribose

14) When two nucleotides are linked together, between which two components is the phosphate ester linkage formed? Using the ultrashorthand forms, diagram a dinucleotide composed of adenosine monophosphate and guanosine monophosphate (use D in both nucleotides.)

–P and (D or R)

–P—D—P—D
      | |
      A   G

15) Write a shorthand form diagram of a chain of 5 nucleotides [adenosine monophosphate (AMP), cytidine monophosphate (CMP), GMP, AMP, and TMP]. Use:

P = Phosphate	A = adenine	G = guanine
D = Deoxyribose	T = Thymine	C = Cytosine

Circle each nucleotide.

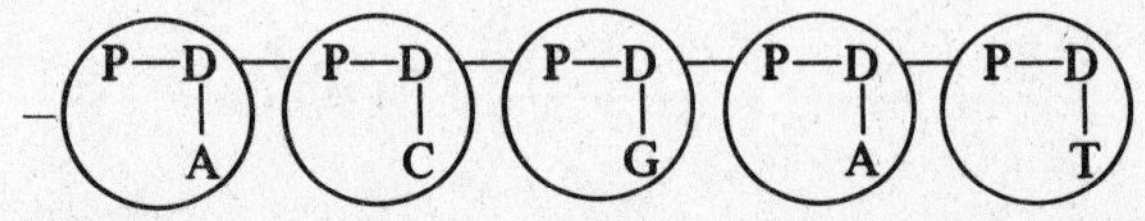

You have diagrammed a short segment of *deoxyribose nucleic acid – DNA. Genes* are constituted of *DNA*.

⑯ In deoxyribose nucleic acid (DNA), the pentose sugar is always ________? DNA is composed of four types of nucleotides, each containing the N base adenine or cytosine or guanine or thymine. Which of these N bases are purines and which are pyrimidines?

–Deoxyribose

–Adenine, Guanine } purines Thymine, Cytosine } pyrimidines

17) Write the shorthand form for a DNA strand containing the following sequence of six nucleotides: GMP, CMP, TMP, CMP, AMP, TMP.

–P—D—P—D—P—D—P—D—P—D—P—D
      | | | | | |
      G   C   T   C   A   T

Note: the P–D–P–D etc. chain is frequently called the *backbone*.

CONCEPTS: NUCLEIC ACIDS

DNA is nucleic acid made up of deoxyribose nucleotides. DNA is the carrier of genetic information in the cell. DNA is the "backbone" of chromosomes. DNA is made up of a A, G, T, and C nucleotides, but not U.

DNA is usually double stranded.

The two strands are held together by H bonds involved in specific N-base pairing:

Double stranded DNA is coiled in a helix structure which was first suggested by Watson and Crick in 1953:

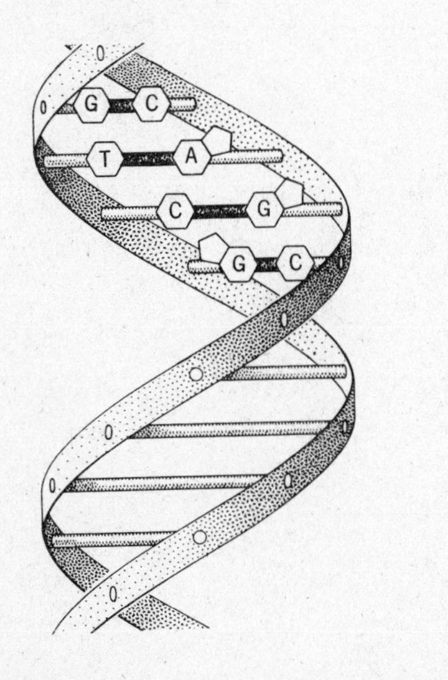

Shorthand notation for a penta nucleotide double strand DNA:

```
P
|
D—A ::: T—D
|         |
P         P
|         |
D—C ::: G—D
|         |
P         P
|         |
D—T ::: A—D
|         |
P         P
|         |
D—C ::: G—D
|         |
P         P
|         |
D—G ::: C—D
          |
          P
```

RNA is nucleic acid made up of ribose nucleotides. RNAs function as genetic messengers, carrying information from DNA to the biochemical "machinery" of the cell. RNAs also function as components of ribosomes and in control of protein synthesis. RNA contains U instead of T and may also have N bases in its makeup other than those listed here. RNA is usually single stranded.

QUESTIONS AND PROBLEMS

SAMPLE RESPONSES

18) The DNA molecule consists of two strands of DNA that are linked by H-bonding between the N-bases: (A and T) and (G and C), thus:

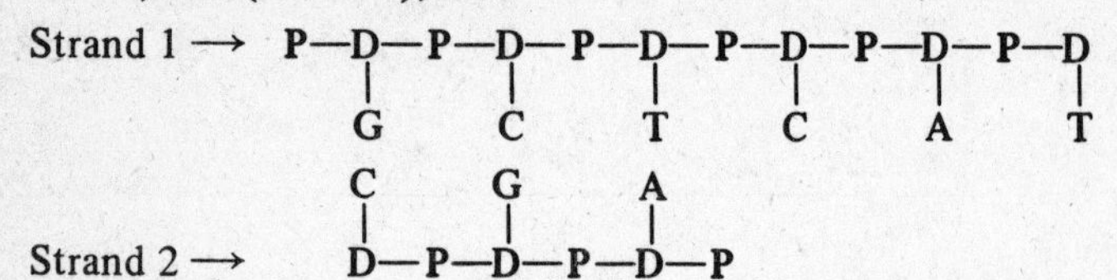

Complete strand 2. Lable the backbone of each strand.

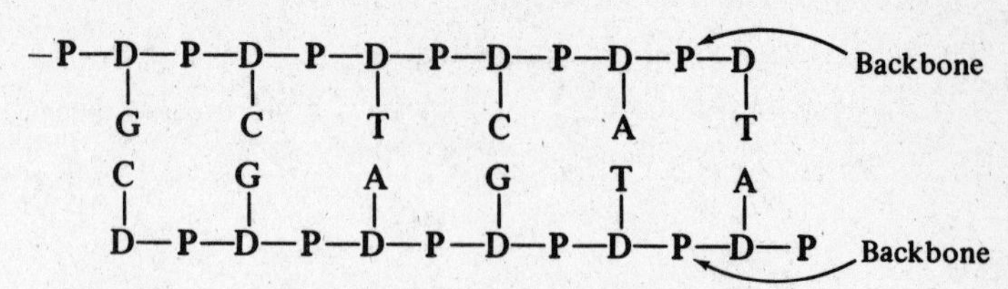

—(Note that the strands "run" in opposite directions.)

19) List the four N-bases that occur in DNA. Indicate the N-base pairing that occurs in DNA structure.

—Adenine, Thymine } pair   Cytosine, Guanine } pair
(Note that the C and G, which look alike, pair.)

20) Write in the H bonds that could form between thymine and adenine.

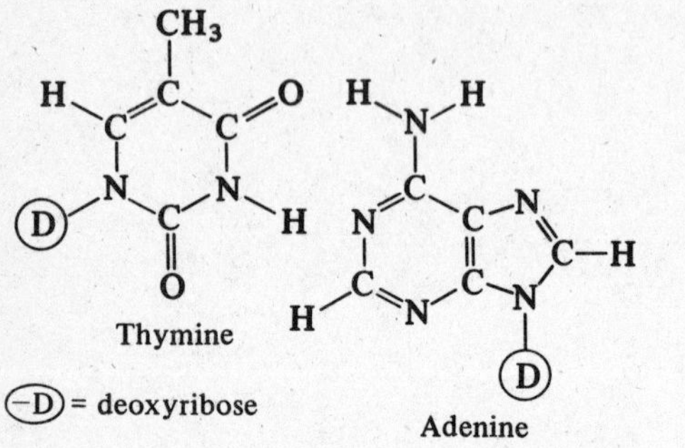

Thymine

(D) = deoxyribose

Adenine

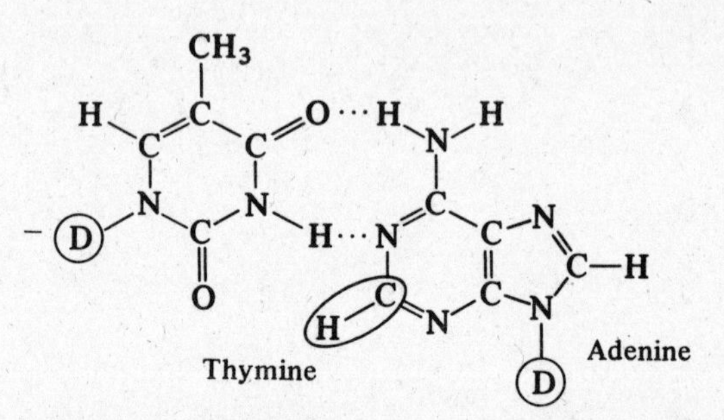

These two H bonds are responsible for A—T pairing. Remember that the H of C—H (circle) does *not* form H bonds.

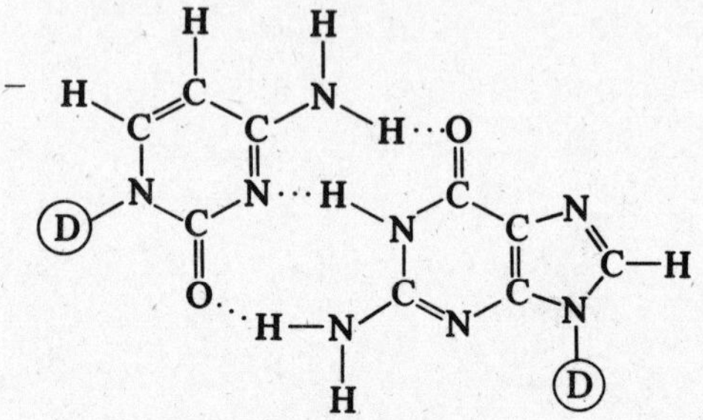

21) Write in the H bonds that could form between cytosine and guanine.

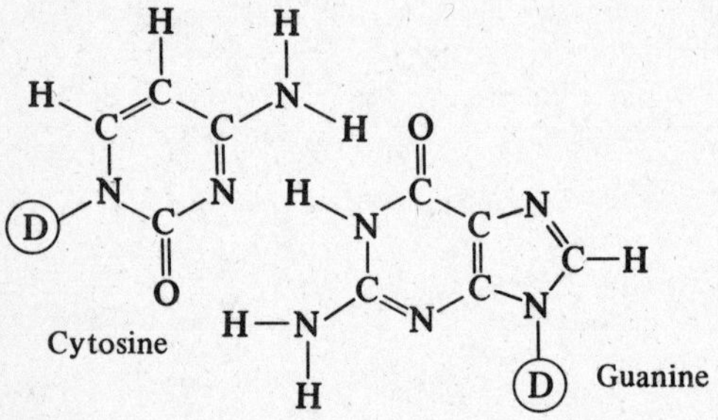

Cytosine

Guanine

These three H bonds are responsible for C—G pairing.

22) How many H bonds are formed between each of the following N-base pairs?

A—T____   G—C____

—A—T_2_ (A:::T)   G—C_3_ (G⋮⋮⋮C)

(23) Write the complementary half of this DNA chain. Include all H bonds (:::) or (⋮⋮⋮). Circle each nucleotide in the entire molecule.

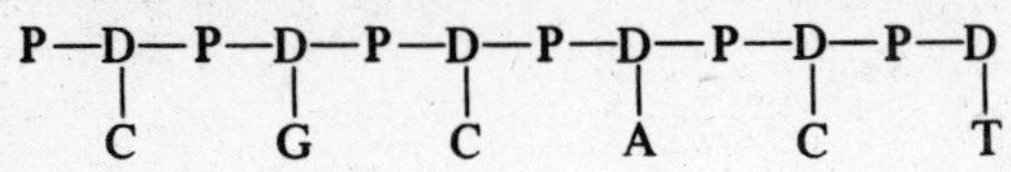

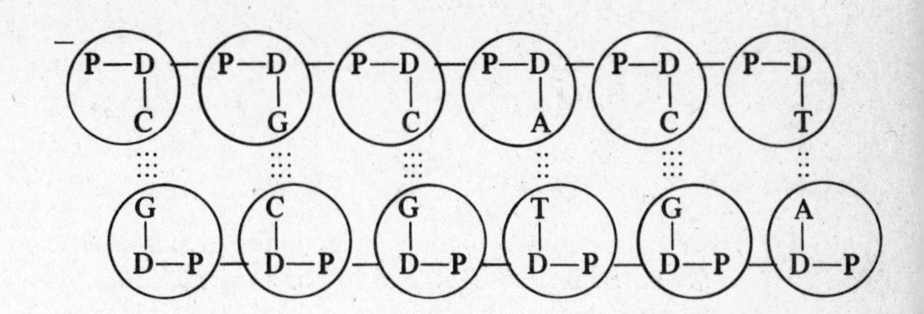

24) This double-stranded structure is coiled in a helical configuration as is illustrated by the diagram on the Concepts page. Who first postulated this structure?

–Watson and Crick (in 1953)

25) RNA = *ribose nucleic acid.* T does not occur in RNA (T is replaced by U = uracil). Write the *RNA* complement to this *DNA* strand. Include all H bonds.

P—D—P—D—P—D—P—D—P—D—P—D ← DNA strand

A T C A T G

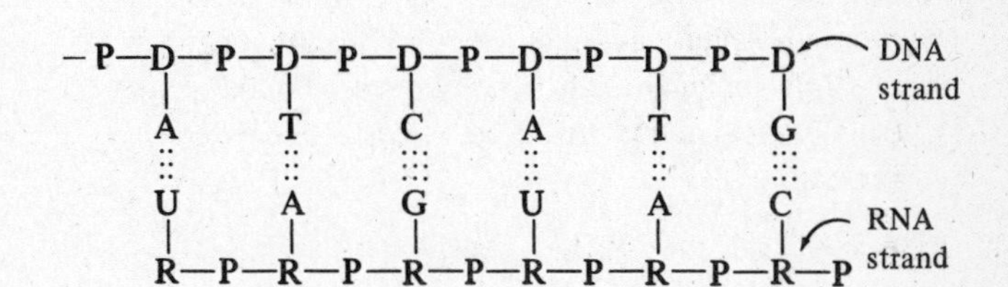

(26) RNA *does not* (usually) form a double-stranded helix. RNA can be manufactured in the cell from genetic DNA. Construct the RNA complement to this genetic DNA strand. Circle each nucleotide.

P—D—P—D—P—D—P—D—P—D—P—D—P—D

A T G A C C T

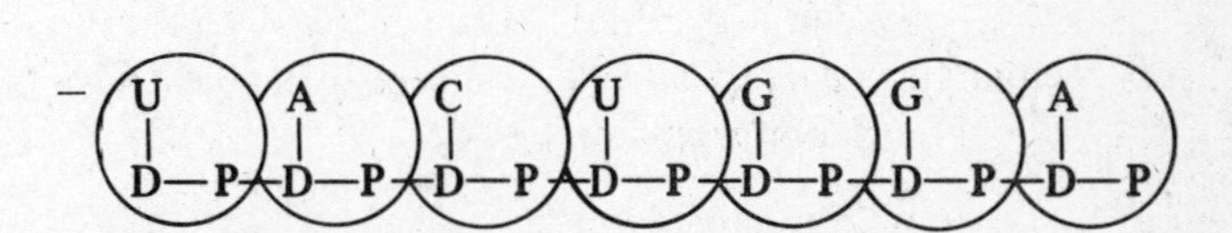

(27) List three differences between DNA and RNA.

–Uracil (RNA) replaces thymine (DNA).
Ribose (RNA) replaces deoxyribose (DNA).
DNA usually occurs in the *double-stranded helix*. RNA is usually *single stranded*.
(or similar responses) (any order)

(28) What name is applied to the model for the molecular structure of DNA?

–Watson-Crick double helix

29) Complete this structure formula for adenine:

NH_2 | C, N, C, N (partial ring structure)

– NH_2 | adenine: N=C–C–N, C–H, N–H, H–C=N (completed purine ring)

30) Write the structural formula for adenine. (Do not consult the previous items.)

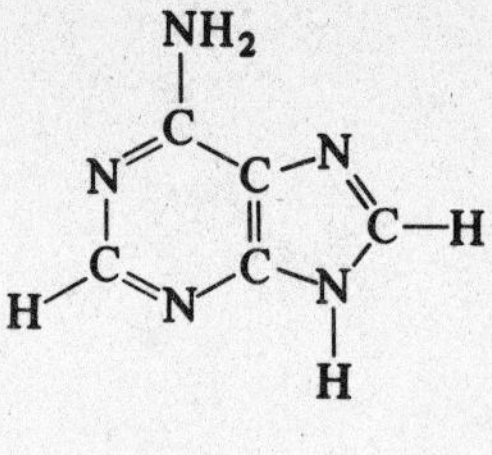

31) Use your previous structural formula for adenine; add to it to diagram the structural formula for adenosine monophosphate (ribose form).

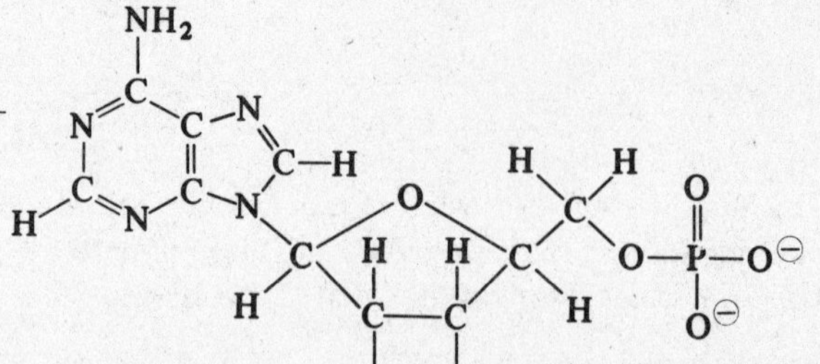

Adenosine monophosphate (AMP)

32) Use the previous structural formula for adenosine monophosphate and add a second phosphate group by bonding via an O to the first phosphate. What other (small) molecule is produced when this linkage is formed?

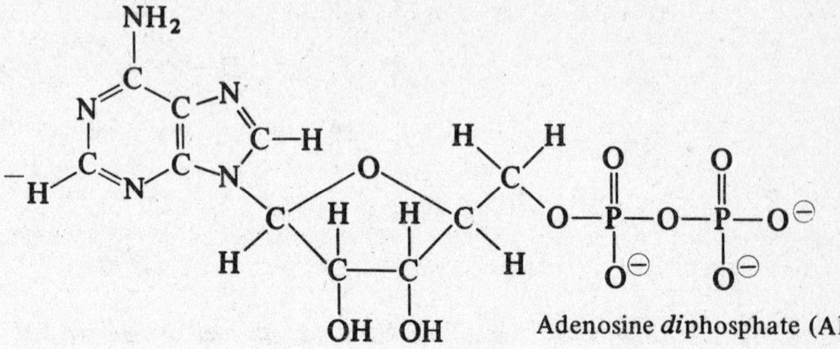

Adenosine *di*phosphate (ADP)

—Water (H_2O) (The 2 H come from H^+ in solution.)

33) Use the previous structural formula for adenosine diphosphate and add a third phosphate group to the second phosphate. What is the name of this molecule? What abbreviation would be used for it?

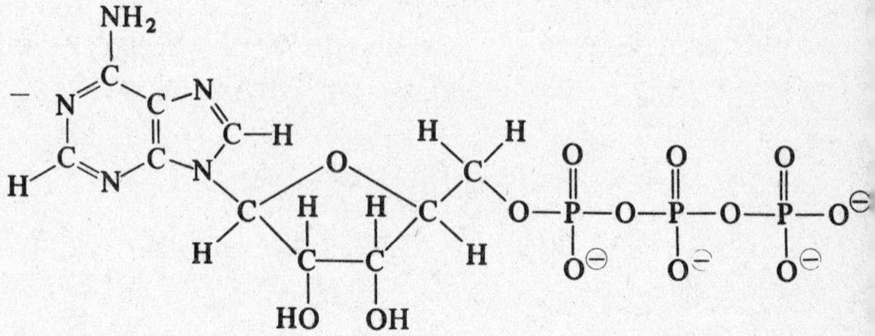

—Adenosine triphosphate (ATP). ATP functions in cellular energy transfer.

34) Write the shorthand structural formula for ATP. What is the chief function of ATP?

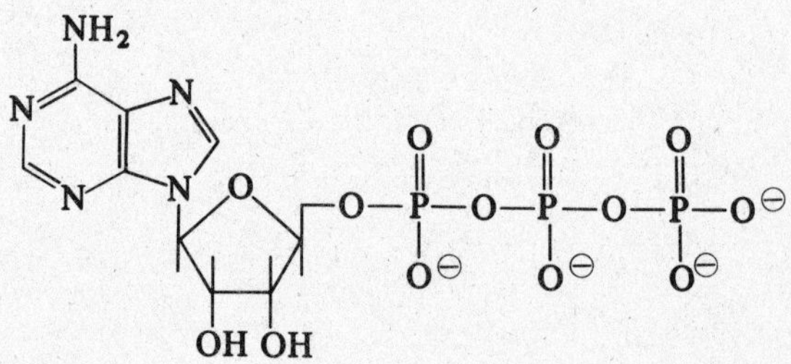

35) The divalent cations Ca^{++} and Mg^{++} bind readily to ⊖ ionic groups.
Ca^{++} is present in most cells only at *very* low concentrations (10^{-6} to 10^{-9} M).
Mg^{++} is present at relatively high concentrations (1 to 10 mM).
Diagram the binding between 2 Mg^{++} and the phosphates of an ATP molecule.

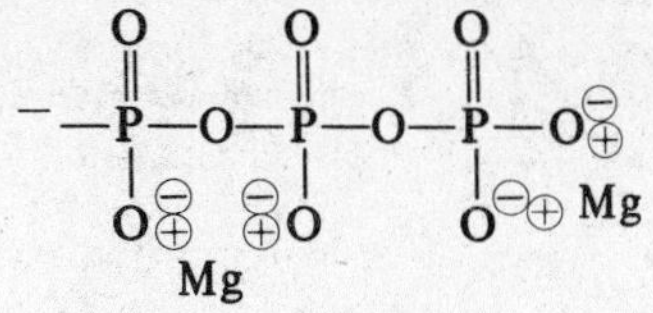

$$-\overset{\overset{\displaystyle O}{\|}}{\underset{\underset{\displaystyle O^{\ominus}}{|}}{P}}-O-\overset{\overset{\displaystyle O}{\|}}{\underset{\underset{\displaystyle O^{\ominus}}{|}}{P}}-O-\overset{\overset{\displaystyle O}{\|}}{\underset{\underset{\displaystyle O^{\ominus}}{|}}{P}}-O^{\ominus}$$

(Two $Mg^{\oplus}$ shown bound: one bridging the $O^{\ominus}$ of the first two phosphates, one bridging the two $O^{\ominus}$ of the terminal phosphate.)

(Virtually all ATP is bound to Mg^{++} in living cells.)

36) Complete the following equation:

ADP + P ⇌ ______ + H_2O

–ADP + P ⇌ ATP + H_2O

37) Complete the following equations:

AMP + P ⇌ ______ + ______

ATP + H_2O ⇌ ______ + P

AMP + 2P ⇌ ______ + ______

–AMP + P ⇌ ADP + H_2O
–ATP + H_2O ⇌ ADP + P
–AMP + 2P ⇌ ATP + $2H_2O$

38) What is the chief characteristic of a basic protein? (If you do not remember for sure, guess.)

–A basic protein contains a high proportion of the basic (NH_2) amino acids, lysine and arginine. (or similar response)

39) What ionic groups would be particularly involved in the bonding between a histone and the *backbone* of a nucleic acid?

–Histone: ⊕ charged groups of lysine, arginine, and histidine.

Nucleic acids: ⊖ from $-\overset{\overset{\displaystyle O}{\|}}{\underset{\underset{\displaystyle O^{\ominus}}{|}}{P}}-$

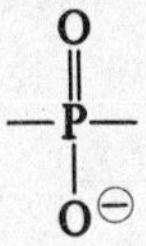

(40) What are the structural units of the nucleic acids?

–Nucleotides

(41) Write out the full names for DNA and RNA.

–DNA = deoxyribosenucleic acid
RNA = ribose nucleic acid

Take at least a 5-minute break before continuing on to the next unit.

SELF-TEST UNITS 14 & 15

(Criterion score 96 points: total of 101 points)

(5 points) 1) Diagram the structural formula for a fat composed of stearic acid: $C_{17}H_{35}COOH$.

(5 points) 2) Write a shorthand diagram for the structure of a phospholipid. Label each component.

(3 points) 3) Write an ultrashorthand diagram for the structure of a phospholipid. Label the diagram appropriately to indicate solubility in water.

(7 points) 4) Diagram the structure of a biological membrane and label the diagram to indicate its constituents.

(5 points) 5) Write an outline structural formula for a sterol. Label the hydrophilic and hydrophobic regions of the molecule.

(2 points) 6) Describe the major functions of cholesterol and cholic acid.

(8 points) 7) List the three types of steroid sex hormones. Describe the major functions of each.

(4 points) 8) Name two adrenal corticosteroid hormones and describe the major function of each.

(8 points) 9) Describe the structure and functions of prostaglandins.

(7 points) 10) Explain the component structure of a nucleotide.

(4 points) 11) List the nitrogen bases of the nucleotides of which DNA is composed.

(5 points) 12) Label each of the following as either a purine or a pyrimidine.

A. [structure: NH_2, N, N, N, N, H]

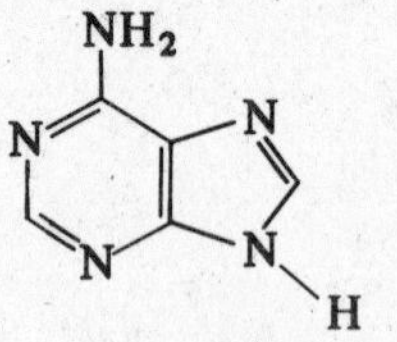

B. [structure: O, H, N, O, N, H]

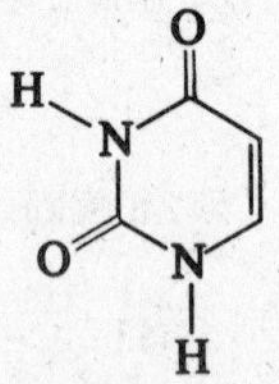

C. [structure: O, H, N, CH_3, O, N, H]

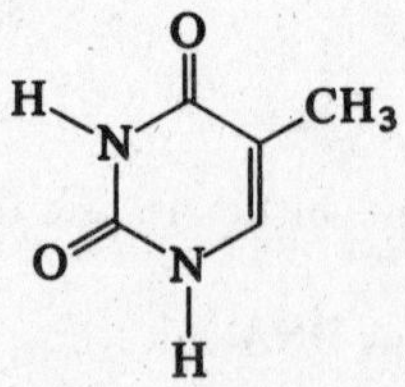

D. Guanine

E. Cytosine

(10 points) 13) Write the complementary strand for this DNA chain: Include all H bonds (:::) or (⁝⁝⁝). Circle each nucleotide in the complementary strand.

P—D—P—D—P—D—P—D—P—D—P—D
C G C A C T

(3 points) 14) What name is applied to the model for the molecular structure of DNA?

(10 points) 15) Construct the RNA complement to this DNA strand: Circle each nucleotide.

P—D—P—D—P—D—P—D—P—D—P—D
A T G A C T

(3 points) 16) Explain the differences between DNA and RNA.

(5 points) 17) Diagram the outline structural formula for ATP.

(4 points) 18) Complete the following equations: ADP + Ph $\rightleftharpoons$ ___ + ___

ATP + H_2O $\rightleftharpoons$ ___ + Ph

AMP + ATP $\rightleftharpoons$ ADP + ___

(1 point) 19) What are the structural units of DNA and RNA?

(2 points) 20) Write out the full names for DNA and RNA.

(101 points total)

SELF-TEST KEY UNITS 14 & 15

(Criterion score 96 points: total of 101 points)

(5 points) 1)

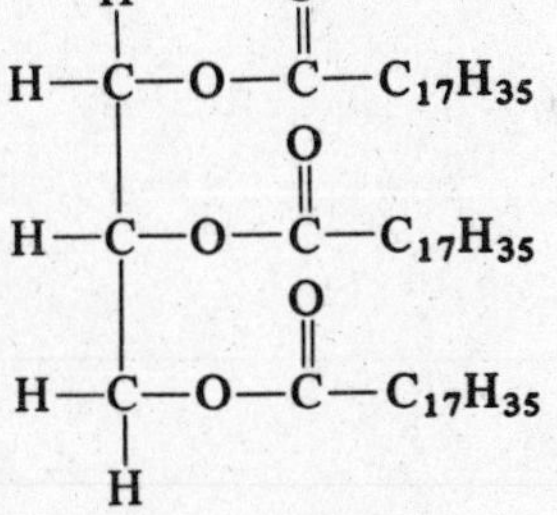

```
      H     O
      |     ‖
  H—C—O—C—C17H35
      |     O
      |     ‖
  H—C—O—C—C17H35
      |     O
      |     ‖
  H—C—O—C—C17H35
      |
      H
```

(stearin–beef fat)

(5 points: 1 point off for each error)
(*Review: Unit 14, Items, 1-11*)

(5 points) 2)

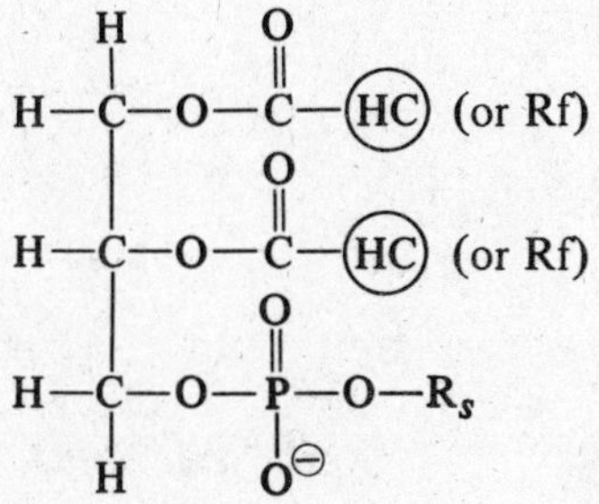

```
      H     O
      |     ‖
  H—C—O—C—(HC) (or Rf)
      |     O
      |     ‖
  H—C—O—C—(HC) (or Rf)
      |     O
      |     ‖
  H—C—O—P—O—Rs
      |     |
      H     O⊖
```

(5 points: 1 point off for each error)
(*Review: Unit 14, Items 11-15*)

(3 points) 3) Hydrophilic / Hydrophobic

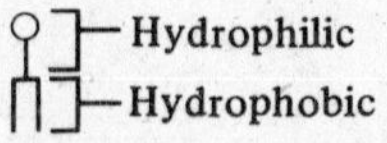

(3 points: 1 point off for each error)
(*Review: Unit 14, Items 12-16*)

(7 points) 4) Phospholipids (1) — About 80% (1)
Lipid Bilayer diagram (2)
Sterols (1) (Cholesterol) (1) — About 20% (1)

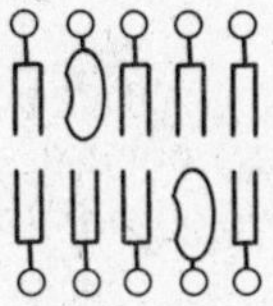

(*Review: Unit 14, Items 16-20 and 27-28*)

(5 points) 5) Hydrophilic → (HO); Hydrophobic

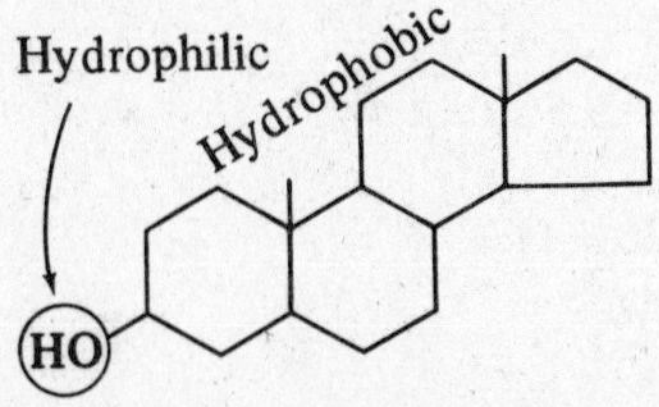

(5 points: 1 point off for each error)
(*Review: Unit 14, Items 21-29*)

(2 points) 6) Cholesterol: major sterol constituent of biological membranes (1).

Cholic acid: sterol detergent in liver bile; facilitates emulsification of fats in digestion (1). (*Review: Unit 14, Items 25-28 and 30-31*)

(8 points) 7) Androgens (testosterone) (1): controls the development of secondary sex characteristics in males (1); for example, beard and hair growth, muscle development, stature, voice tone, and male behavior (1 point for a reasonable listing of characteristics).

Estrogens (estrone) (1): controls the development of female secondary sex characteristics (1); for example, development of mammary glands, subcutaneous fat deposition, pelvic configuration, uterus development, and female behavior (1 point for a reasonable listing of characteristics).

Progesterone (1): regulates uterus development, pregnancy, and the onset of labor (1). (*Review: Unit 14, Items 32-36*)

(4 points) 8) Aldosterone (1): regulates salt balance via the kidney (1).

Corticosterone (1): regulates carbohydrate metabolism in all cells (1).
(Review: Unit 14, Items 37–40)

(8 points) 9) Prostaglandins contain a chain of 20 C atoms (1) with a methyl group at one end (1) and a carboxyl group at the other (1) and with a ring structure in the middle so that the molecule is folded into a hairpin configuration.

Prostaglandins function in intercellular communication (1), intracellular communication (1), regulation of smooth muscle contraction (1)—particularly in blood vessels (1)—and affect the permeability of cell membranes to salts and water (1).
(Review: Unit 14, Items 42-43)

(7 points) 10) Nitrogen base (1), either a purine (1) or pyrimidine (1); plus a pentose sugar (1), either deoxyribose (1) or ribose (1); plus a phosphate group (1). *(Review: Unit 15, Items 1-10)*

(4 points) 11) Adenine, thymine, guanine, and cytosine (1 point each).
(Review: Unit 15, Items 11-17)

(5 points) 12) A. Purine C. Pyrimidine E. Pyrimidine
B. Pyrimidine D. Purine (1 point each)
(Review: Unit 15, Items 11-17)

(10 points) 13)

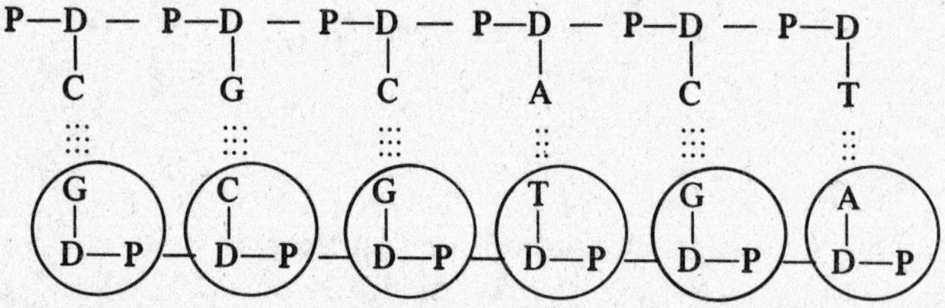

P—D — P—D — P—D — P—D — P—D — P—D
C G C A C T
G C G T G A
D—P D—P D—P D—P D—P D—P

(10 points: 1 point off for each error)
(Review: Unit 15, Items 18–23)

(3 points) 14) Watson- (1) Crick (1) double helix (1) *(Review: Unit 15, Item 24)*

(10 points) 15)

(10 points: 1 point off for each error)
(Review: Unit 15, Items 25–27)

U A C U G A
R—P R—P R—P R—P R—P R—P

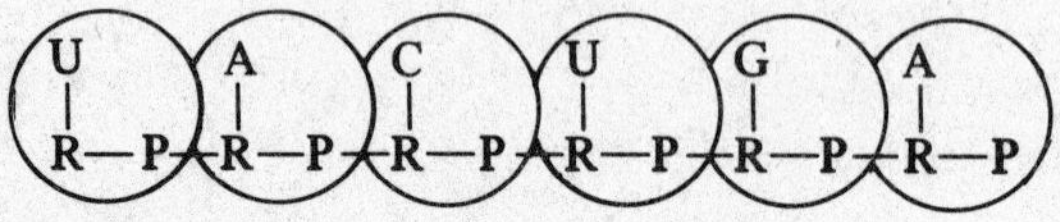

(3 points) 16) RNA contains ribose; DNA contains deoxyribose (1).
In RNA uracil replaces thymine (of DNA) (1).
RNA is usually single stranded; DNA is usually double stranded (1).
(Review: Unit 15, Items 25–27)

(5 points) 17)

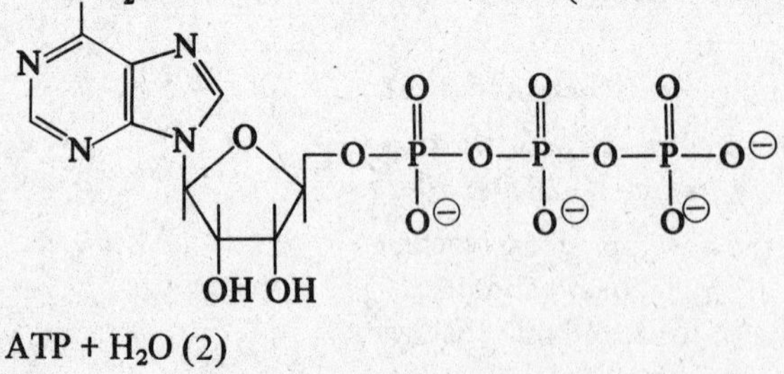

NH$_2$ N N N N O —O—P—O—P—O—P—O$^{\ominus}$ O O O O$^{\ominus}$ O$^{\ominus}$ O$^{\ominus}$ OH OH

(4 points) 18) ATP + H_2O (2)
ADP (1)
ADP (1) *(Review: Unit 15, Items 36-37)*

(1 point) 19) Nucleotides (1) *(Review: Unit 15, Concepts)*

(2 points) 20) Deoxyribose nucleic acid and ribose nucleic acid (1 point each).
(Review: Unit 15, Concepts)

(101 points total)

Take at least a 5-minute break before going on to the next unit.

UNIT 16 — SIXTEEN ENERGY

CONCEPTS

One of the fundamental properties of living systems is their ability to transform energy.

Kinetic energy is the active state of energy; it is the energy of motion; it is energy released.

Potential energy is the inactive state of energy; it is the energy of arrangement or position; it is energy stored.

The greater the degree of order in a system (or molecule), the greater the level of potential energy.

Energy is transferred as the order within molecules is changed (or as the relative complexity of molecules is changed).

Differences between the energy levels of molecules are the sources of energy for all living cells and organisms.

An *exothermic* reaction releases heat (energy).

An *endothermic* reaction takes up or stores heat (energy).

The Calorie is the unit usually used for measuring the amount of energy transfer in biochemistry.

1 calorie = the energy to raise the temperature of 1 gram of water 1°C. (Celsius).

1 Calorie = 1000 calories = 1 kilocalorie

The *energy of activation* is the amount of kinetic energy reactant molecules must possess to undergo a given reaction and become products.

QUESTIONS AND PROBLEMS / SAMPLE RESPONSES

1) Label each of the following as an example of either *potential energy* or *kinetic energy*:

A. A landslide
B. Balancing rock
C. Water (spilling) over the dam
D. Molecular motion
E. The lake of water behind the dam
F. The energy of a covalent bond
G. Hot coffee (as opposed to cold coffee)
H. A chocolate bar

–A. Kinetic
B. Potential
C. Kinetic
D. Kinetic
E. Potential
F. Potential
G. Kinetic ("heat" is increased molecular motion)
H. Potential

② Define briefly potential energy and kinetic energy.

–Potential energy is restrained energy which can be available to do work upon removal of the restraint. (e.g., water behind the dam)
–Kinetic energy is the energy of motion. (e.g., molecular motion)

3) Persons who are overweight are sometimes concerned about the energy content of their diet. In what units do they usually count the energy content of foods?

–Calories

4) Thus, the biologist would write:

Glucose + $6O_2$ ⇌ $6CO_2$ + $6H_2O$ + 673 ___?___ (units of energy)

—Calories

5) In a closed system (e.g., a thermos bottle) it takes:

250 calories of heat to raise 250 ml H_2O 1°C.
37 calories of heat to raise 1 ml H_2O 37°C.
100 calories of heat to raise 10 ml H_2O 10°C.

Define a calorie.

—A calorie is the amount of heat required to raise 1 ml H_2O 1°C. (or similar response)

⑥ 1 Calorie = 1 kilocalorie = 1000 calories

How many Calories does 3000 calories equal? —3

How many calories does 132 Calories equal? —132,000

How many Calories does 132 calories equal? —0.132 $\frac{132}{1000}$

⑦ Which of the following molecules would you expect to contain the greatest potential energy?

A. CH_3—COOH (H—C(H)(H)—C(=O)—OH)

B. $CH_3CH_2CH_2CH_2$—COOH (H—C(H)(H)—C(H)(H)—C(H)(H)—C(H)(H)—C(=O)—OH)

C. H—C(H)(NH_2)—C(=O)—OH

Explain your answer.

—B, because B contains the greatest number of bonds (the greatest level of organization or order).

⑧ Given: H—C(H)(H)—C(=O)—OH + $2O_2$ ⇌ $2CO_2$ + $2H_2O$ + Energy

$6CO_2$ + $6H_2O$ + energy ⇌ glucose + $6O_2$
Pyruvic Acid ⇌ acetic acid + CO_2 + energy

In general: Is energy consumed or released as a large molecule is formed? Is energy consumed or released as a large molecule is broken down?

—Formed: energy is consumed.
—Broken: energy is released.

⑨ An *exothermic* reaction:

Glucose + $6O_2$ ⇌ $6CO_2$ + $6H_2O$ + Energy

An *endothermic* reaction:

$6CO_2$ + $6H_2O$ + Energy ⇌ Glucose + $6O_2$

From what you can reason out about the word meanings, define an exothermic and an endothermic reaction.

—An *exothermic* reaction *releases* energy (heat).
—An *endothermic* reaction *consumes* energy. (or similar responses)

(10) Label each of the following reactions as either exothermic or endothermic. (If you are in doubt check back to Item No. 8.)

A. Pyruvic acid + $3O_2 \rightleftharpoons 3CO_2 + 3H_2O$
B. $C_4H_6O_4 \rightleftharpoons C_4H_4O_4 + 2(H)$
C. Amino acids $\rightleftharpoons$ polypeptide chain (protein)
D. ADP + P $\rightleftharpoons$ ATP
E. Glycerol + 3 fatty acids $\rightleftharpoons$ fat

–A. Exothermic
B. Exothermic
C. Endothermic
D. Endothermic
E. Endothermic

11) In a collection of molecules, for example, let us take a solution of ATP in water, all molecules are not moving at the same rate; therefore, they have different kinetic energy levels. This can be illustrated by the plot:

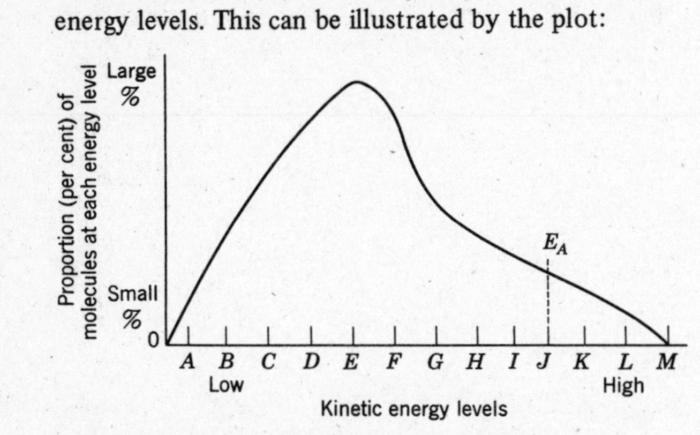

Which *energy level* is higher (*D* or *G*)? Which energy level contains the greatest number of molecules of ATP (*D* or *G*)? Would the number of molecules of ATP at energy level *J* or higher be relatively large or small?

–*G* (*A* is low; *M* is high)

–*D*

–Small

12) In order for two molecules to combine, they must come together or collide. Which would facilitate collisions, a high or low kinetic energy? In the reaction ATP + $H_2O \rightleftharpoons$ ADP + Ph + 10 Cal.,* which two molecules must collide for ADP and Ph to be produced?

–A high kinetic energy (will facilitate collisions).

–ATP + H_2O

*The amount of energy change involved in the reaction ADP + Ph $\rightleftharpoons$ ATP + H_2O is dependent on the conditions under which the reaction occurs.

Under "Standard Conditions" of 1 atmosphere of pressure, 23°C, and one molar concentrations of reactants the energy change is about 7 Calories per mole of ATP produced.

Under *physiological* conditions in living cells the temperature may vary from roughly 0°C to 40°C, the pressure may be greater than 1 atm because of surface forces in small cellular organelles and the concentrations of the reactants are variable but in the range of 0 to 50 *milli*moles. In addition, the energy change is influenced by the concentrations of ions in the cellular environment, particularly that of Mg^{++}.

As a result of these factors the energy change of the reaction may vary from 7 to 12 Calories per mole of ATP produced. As a generalized simplification we will use a value of 10 Cal per mole ATP produced under physiological conditions.

(13) Not all molecules of ATP and H_2O that collide can react. In the reaction ATP + H_2O $\rightleftharpoons$ ADP + Ph + 10 Cal, the bond between the second Ph and the third Ph of ATP must be broken if the reaction goes to the right. Only a few colliding molecules of ATP and H_2O have enough kinetic energy to allow this bond to be broken. This energy level is called the energy of activation. In the plot of item 11, the energy of activation (E_A) is labeled. Is the proportion of molecules that have a kinetic energy level above E_A large or small? In your own words, define the energy of activation.

–Small
–The energy of activation is the minimum energy level that two colliding molecules must possess in order to undergo a given chemical reaction. (or similar response)

14) Given the reaction: ATP + H_2O $\rightleftharpoons$ ADP + Ph + 10 Cal, the following relationship holds:

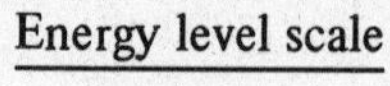

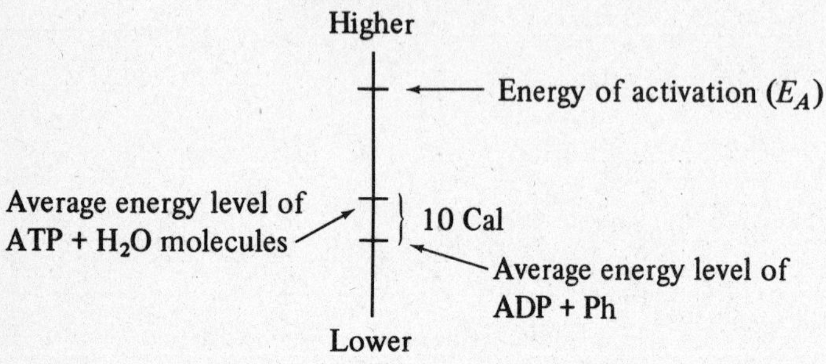

For which reaction (right or left) is the activation energy the greatest? Which group of molecules (ATP + H_2O or ADP + Ph) would you then expect to have the greatest proportion of molecules at a kinetic energy level above E_A?

–Left (ADP + Ph are the reactants.)

–ATP + H_2O

(15) Define chemical equilibrium for the reaction: ATP + H_2O $\rightleftharpoons$ ADP + P + 10 Cal

–At equilibrium, rate to the right = rate to left.

(16) If ATP is added to the reaction at equilibrium, what will be the initial effect on the rate of the reaction to the right?

In general:

An increase in the concentration of the reactants will cause a(n) _____ in the rate of the reaction to the right.

–Increase

An increase in the concentration of the products will cause a(n) _____ in the rate of the reaction to the left.

–Increase

A decrease in the concentration of the reactants will cause a(n) _____ in the rate of the reaction to the right.

–Decrease

A decrease in the concentration of the products will cause a(n) _____ in the rate of the reaction to the left.

–Decrease

17) The rate of a reaction is determined by the number of molecules that are at an energy level above the energy of activation. At equilibrium, what relationship must be true for the number of molecule pairs of ATP + H_2O and ADP + Ph that are above E_A?

–(ATP + H_2O) = (ADP + Ph) (i.e. the rate to the right = the rate to the left.)

18) ATP + H_2O $\rightleftharpoons$ ADP + Ph + 10 Cal. If the reaction to the left has a higher energy of activation than the reaction to the right (see item 14), will a higher or lower concentration of ADP + P (as compared to the concentration of ATP + H_2O) be present at equilibrium?

–Higher (The higher the E_A the lower the proportion of molecules above E_A.)

19) At equilibrium, which would be present in higher concentration ATP or ADP?

–ADP

(20) $CH_4 + 2O_2 \rightleftharpoons CO_2 + 2H_2O$ + 213 Cal. Which reaction (right or left) has the greatest E_A? At equilibrium, would CH_4 or CO_2 be present in the greatest concentration?

–Left

–CO_2

(21) $H_2 + I_2$ + 230 cal $\rightleftharpoons$ 2HI (I = iodine). At equilibrium, which is present in higher concentration, H_2 or HI?

–H_2

(22) A catalyst lowers the activation energy for a reaction. In the plot of item 11, would a catalyst shift E_A to the right or left? Would a catalyst affect the rate of the reaction to the right? Would a catalyst affect the rate of the reaction to the left? The effects of a catalyst on the rates to the right and left are exactly the same. Would a catalyst affect the equilibrium concentrations of the reactants and products?

–Left
–Yes
–Yes

–No (a catalyst alters the rate at which a reaction reaches equilibrium–but not the equilibrium concentrations.)

23) An enzyme is a biological catalyst. What is the effect of an enzyme on a chemical reaction?

–An enzyme lowers the energy of activation for the reaction. (It increases the right and left rates equally.) (or similar response)

24) Would an enzyme facilitate (speed up) the attainment of equilibrium? Explain your answer.

–Yes. An enzyme increases the rate of both the right and left reactions, thus facilitating the attainment of equilibrium.

25) Enzymes act by combining with or binding the reacting (*substrate*) molecules. Which group of biochemical compounds you have studied (carbohydrates, proteins, lipids, or nucleic acids) has the greatest general capacity for reacting with other molecules?

–Proteins. (Indeed, *all enzymes are proteins.*)

Take at least a 5-minute break before continuing on to the next unit.

UNIT 17 — SEVENTEEN ENZYMES

CONCEPTS

Almost all chemical reactions that occur in living cells are facilitated by enzymes.

An enzyme increases the rate at which a specific reaction reaches equilibrium. It does not change the equilibrium point.

All enzymes are proteins.

Many, probably most, cellular enzymes are bound to membranes. Membrane bound enzymes can function as integrated enzyme teams, which facilitate sequential chemical reactions (metabolic pathways).

The molecules that are acted upon by enzymes are termed substrates.

Enzymes function by binding molecules in a specific configuration at binding sites on the enzyme molecule. Because of specific conformational binding, enzymes facilitate the formation or breaking of bonds between substrates.

Binding of Reactants and Products to an Enzyme

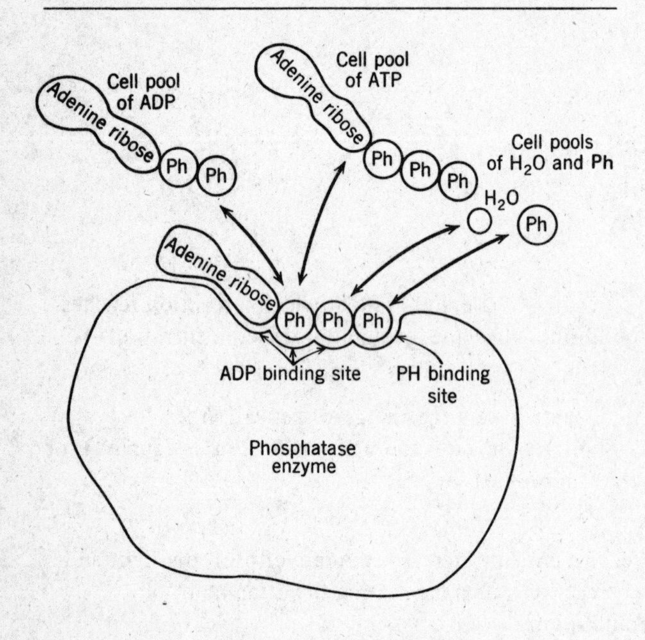

QUESTIONS AND PROBLEMS

SAMPLE RESPONSES

1) Write an equation that would be facilitated by the phosphatase enzyme in the concepts diagram.

–ADP + Ph + energy $\rightleftarrows$ ATP + H_2O

2) In the preceding equation, can you tell which molecules are reactants and which are products? Explain your response.

–No. It depends on the point of view. Viewed from the left to right ADP and Ph are reactants, while from right to left ATP and H_2O are reactants. This must be defined for each reaction. (or similar response)

3) In the reaction:

$$CO_2 + H_2O \rightleftharpoons H_2CO_3 \text{ (carbonic acid)}$$

If the reacting molecules are at equilibrium, what would be the effect of adding CO_2 to the system?

–Adding CO_2 increases the CO_2 concentration, and the reaction rate toward the right would increase.

4) If the carbonic acid reaction is at equilibrium, what would be the effect of removing CO_2?

–The overall reaction rate toward the left would increase. (The reaction rate toward the right would slow down– and the reaction rate to the left would predominate.)

5) Glucose + ATP $\xrightarrow[\text{Hexokinase}]{}$ Glucose-1-Ph + ADP

This reaction occurs in cells. Assume that the reaction is proceeding slowly toward the right. If the cell then takes in more glucose will the reaction speed up or slow down? Toward the right or left?

–Speed up toward the right

6) What are molecules that are bound to an enzyme called?

–Substrates

7) Relatively weak, short-range bonds hold the substrate molecules on the enzyme binding sites.

There is a rapid turnover–usually hundreds or thousands of molecules bound and released per second:

bind-release-bind-release-bind-release-etc.

An enzyme can bind either reactants or products. In the example: either ADP and Ph or ATP and H_2O. If a molecule of phosphatase has bound a molecule of ADP and Ph, can it also bind a molecule of ATP at the same time? Explain your response.

–No

–The binding sites are already occupied by the ADP and Ph. ATP could be bound only after the ADP and Ph are released.

8) If Ph is bound to phosphatase, can the enzyme molecule bind a molecule of ATP?

–No. To bind ATP the Ph binding site must be free to bind the 3rd Ph of ATP.

Can it bind a molecule of ADP?

–Yes. The entire ADP binding site is free.

9) Look at the diagram in concepts:

Phosphatase can facilitate the reaction in *either* direction. Substrate can be bound as: ATP *or as* ADP + Ph and released as ATP *or as* ADP + Ph.

If the ADP and Ph concentrations are high and the ATP concentration is low, which molecules are most likely to bind to phosphatase?

–ADP and Ph

10) If the ADP and Ph concentrations are high and the ATP concentration is low, which molecules are most likely to be released from the enzyme?

–ATP

(11) Based on your understanding of enzyme action, how do you think an enzyme lowers the energy of activation for a reaction?

–The enzyme binds the substrate molecules in close physical contact, thus facilitating bonding (either making or breaking of the bond). (or similar response)

12) Many enzymes contain nonprotein metal containing groups (e.g., Fe and Mg). What general term would you apply to these metal-containing groups? What would you call a protein that contains such a group?

–Prosthetic group
–A conjugated protein

13) Any single cell or organelle contains a specific number of molecules of any one type. It is convenient to think of these as molecular pools. For example, the ATP pool, ADP pool, NAD^+ pool, etc.

ATP pool + ADP pool = a constant

As ATP is formed from ADP + Ph, what is the effect on the ATP pool? On the ADP pool?

–The ATP pool increases.
–The ADP pool decreases.

14) If the ATP pool is very large, what effect would this have on the ADP pool?

–The ADP pool would be relatively small.

Under these conditions, would ADP or ATP most probably bind to a molecule of phosphatase?

–ATP

Would ATP or ADP most probably be released from a molecule of phosphatase to which ATP has been bound?

–ADP (because the ATP concentration is high and the ADP concentration is low).

15) Assume that the total ATP + ADP pools of a cell is 10,000,000 molecules. If the ADP pool is 2,000,000, what is the ATP pool?

–8,000,000 molecules

If the ADP pool is increased to 3,000,000 by the action of phosphatase, what is the new ATP pool?

–7,000,000 molecules

16) This reaction takes place in the chloroplasts of green plants:

$$\text{Ribulose-1-Ph + ATP} \underset{\text{phosphoribulokinase}}{\rightleftharpoons} \text{ribulose-1-5DiPh + ADP}$$

In the light, chloroplasts change ADP + Ph to ATP + H_2O.

In light, what are the relative sizes of the ADP and ATP pools?

–The ATP pool is large; the ADP pool is small.

In light, would the phosphoribulokinase reaction produce ribulose Ph or ribulose-di-Ph? Explain your response.

–Ribulose-di-Ph.
–In light, the ATP pool is large and the rate of the phosphoribulokinase reaction to the right is increased, forming ribulose-di-Ph. (or similar response)

17) Review: List various treatments that will denature protein molecules.

–Heat, salts, pH, organic solvents, heavy metal ions

18) What effect would you expect the denaturing (with heat, say) of an enzyme to have on its ability to facilitate a reaction?

–Denaturation involves a change in tertiary structure of the protein.
–If the change in tertiary structure altered the configuration of the binding sites, the capability of the enzyme to bind substrate molecules might be altered. (or similar response)

(19) What is an enzyme?

–A protein molecule that facilitates a specific reaction toward equilibrium.

(20) Explain how enzymes function.

–By binding substrate molecules at binding sites, holding the substrates in specific configurations that facilitate the formation or breaking of bonds.

(21) Explain the concept of equilibrium.

–In a chemical reaction, equilibrium is attained when the rate is the same in both directions.
–Each reaction has specific equilibrium concentrations for the reacting molecules.

(22) The enzyme adenyl cyclase is attached to the membranes of many human cells.

$$ATP + H_2O \underset{\text{adenyl cyclase}}{\rightleftharpoons} \text{cyclic AMP} + 2Ph$$

If the ATP pool is low, what effect would this have on the rate of formation of cyclic AMP?

–Cyclic AMP formation would be slow.

(23) This reaction occurs in your blood and body fluids:

$$CO_2 + H_2O \underset{\text{anhydrase}}{\overset{\text{carbonic}}{\rightleftharpoons}} H_2CO_3 \rightleftharpoons H^+ + HCO_3^- \text{ (bicarbonate)}$$

When blood arrives in the lung, CO_2 is lost and the CO_2 concentration decreases.
What is the effect on the reaction facilitated by carbonic anhydrase?

–Its rate increases to the left.

What is the effect on the concentration of H_2CO_3?

–As the reaction proceeds to the left, the concentration of H_2CO_3 decreases.

What is the subsequent effect on the reaction that produces bicarbonate ions?

–The reaction rate increases toward the left.

What is the effect on the bicarbonate concentration of the blood?

–The bicarbonate concentration of the blood decreases.

24) In the system of the preceding question, what is the effect of CO_2 and HCO_3^- loss in the lung on the pH of the blood?

–The pH increases. (H^+ concentration decreases as the reactions "go" to the left.)

Take at least a 5-minute break before beginning the Self-Test.

SELF-TEST UNITS 16 & 17

(Criterion score 38 points: total of 41 points)

(3 points) 1) Explain the difference between potential energy and kinetic energy.

(3 points) 2) What is a calorie?

(1 point) 3) How many Calories are equivalent to 643 calories?

(2 points) 4) Which of the following molecules would you expect to contain the greatest amount of potential energy?
A. $C_{17}H_{35}COOH$ B. $C_{16}H_{33}COOH$ C. $C_{15}H_{31}COOH$
Explain your response.

(2 points) 5) Using a phrase for each, identify the difference between an endothermic reaction and an exothermic reaction.

(5 points) 6) Identify each of the following as exothermic or endothermic:
A. ADP + Ph $\longrightarrow$ ATP + H_2O
B. Glucose + $6O_2 \longrightarrow 6CO_2 + 6H_2O$
C. n glucose $\longrightarrow$ glycogen + nH_2O
D. Pyruvate $\longrightarrow$ Ethyl alcohol + CO_2
E. FructosePh + Ph $\longrightarrow$ Fructose-di-Ph

(3 points) 7) Briefly explain the energy of activation for a reaction.

(2 points) 8) What is the approximate energy change per mole ATP in the reaction:
ATP + $H_2O \rightleftarrows$ ADP + Ph under standard conditions?
. . . under physiological conditions?

(3 points) 9) Pyruvate + NADH + $H^+ \rightleftarrows$ Lactate + NAD^+ + 6.0 Cal/mole
At equilibrium, is pyruvate or lactate present in the greater concentration?
Briefly explain your response.

(3 points) 10) Explain the effects of a catalyst on a chemical reaction.

(2 points) 11) What is an enzyme?

(6 points) 12) Explain how enzymes function.

(4 points) 13) If this reaction is at equilibrium:

$$\text{Pyruvate} + \text{NADH} + H^+ \rightleftarrows \text{lactate} + NAD^+$$

A. What effect would an increase in pyruvate concentration have on the rate?
B. What effect would a decrease in NADH concentration have on the rate?
C. What effect would a decrease in NAD^+ concentration have on the rate?
D. What effect would a decrease in pH have on the rate?

(2 points) 14) In the reaction: NAD^+ + malate $\rightleftarrows$ NADH + oxaloacetate
The NAD^+ pool + NADH pool = a constant.
What is the relative size of the NAD^+ pool if the NADH pool is large?
What is the effect of increasing the NADH pool on the rate of the reaction to the right?

(41 points total)

SELF-TEST KEY UNITS 16 & 17

(Criterion score 38 points: total of 41 points)

(3 points) 1) Potential energy is stored or restrained energy (1) that can be made available to do work (1). Kinetic energy is the energy of motion. (1) *(Review: Unit 16, Items 1-2)*

(3 points) 2) A calorie is the amount of heat required to raise 1 gram of water 1 degree Celsius. (3) *(Review: Unit 16, Items 3-4)*

(1 point) 3) 0.643 (1) *(Review: Unit 16, Items 5-6)*

(2 points) 4) A (1)
Molecule A contains a greater degree of order or organization (1) (a greater number of covalent bonds). *(Review: Unit 16, Item 7)*

(2 points) 5) Endothermic: takes up energy (heat). (1)
Exothermic: releases energy. (1) *(Review: Unit 16, Items 8-9)*

(5 points) 6) A. Endothermic (1 point each)
B. Exothermic
C. Endothermic
D. Exothermic
E. Endothermic *(Review: Unit 16, Item 10)*

(3 points) 7) The energy of activation is the combined minimum energy (1) that reacting molecules must possess (1) to be able to undergo a specific chemical reaction (1). *(Review: Unit 16, Items 11-14)*

(2 points) 8) Standard conditions: 7 Calories/mole (1)
Physiological conditions: 10 Calories/mole (1) *(Review: Unit 16, Item 12)*

(3 points) 9) Lactate (1)
The reaction to the right is exothermic. Hence, the energy of activation for the reaction to the left is greater, and the reactants for the reaction to the left will be present in higher concentration than the reactants for the reaction to the right. (2)
(Review: Unit 16, Items 17-21)

(3 points) 10) A catalyst increases the rate (1) at which a reaction attains equilibrium (1), but does not change the equilibrium concentrations of the reactants and products (1). *(Review: Unit 16, Item 21)*

(2 points) 11) An enzyme is a specific protein molecule (1) that acts as a biological catalyst (facilitating a specific biological reaction) (1).
(Review: Unit 16, Items 23-25)

(6 points) 12) Enzymes function by binding (1) substrate molecules (1) in a specific configuration (1) that facilitates the formation or the breaking of bonds (1).
Enzymes facilitate the rate at which a reaction reaches equilibrium. (1) They do not change the equilibrium concentrations of the reacting molecules. (1) *(Review: Unit 17, Items 1-10)*

(4 points) 13) A. Rate increases to the right. (1 point each)
B. Rate increases to the left.
C. Rate increases to the right.
D. Rate increases to the right. *(Review: Unit 17, Items 1-14)*

(2 points) 14) Small (1)
Decrease to the right (1) *(Review: Unit 17, Items 11-14)*

(41 points total)

CUMULATIVE SELF-TEST

This test can be used both as a pretest, to determine whether a student may omit certain units, and as a posttest to assess learning after the text has been completed. The cumulative test is selective; that is, it does not cover all the material that is stressed by the book. For a thorough self-testing or review the unit self-tests should be used (see Contents).

In this cumulative test, the unit containing the material covered by each question is listed at the left to facilitate pretest selection of items.

Evaluation of responses will require judgment and should be conservative, that is, if there is a question about the adequacy of a response, it should be scored as incorrect. Emphasis should be placed on the correctness of concepts rather than upon the identity of words or phrases.

For pretest use, locate the appropriate items by scanning the unit notations at the left of the test questions. Answer all questions for the given unit you wish to check. For example, if you feel that you probably know the material in Unit 1, answer questions 1 and 2. Check your responses against the key, following the instructions at the beginning of the key. If you judge your responses to be correct, you may omit that unit. If you are in doubt about your answers, you should probably work through the material in the appropriate unit to be certain of your comprehension.

UNIT	QUESTION
1	1. Write the outer shell diagram for the magnesium atom (atomic number = 12).
1	2. Write the structural formula for ($CH_3CHOHCOOH$).
2	3. Match the following:

_____Aldehyde

_____Aromatic

_____Ester linkage

_____Keto

_____Ethyl

A. H—C—C— (each C bonded to H above and H below)

B. —(benzene ring)

C. —C(=O)—O—

D. —C—C(=O)—C—

E. —C(=O)OH

F. —C—H (C bonded to H above and H below)

G. —C(=O)H

H. —N(H)H

I. —O—P(=O)(OH)—OH

J. —OH

3 4. Write the structural formula for sodium acetate. Indicate the ionic bond with an arrow.

$$\text{acetic acid} = \mathrm{H{-}\overset{\displaystyle H}{\underset{\displaystyle H}{\overset{|}{\underset{|}{C}}}}{-}C{\overset{\displaystyle O}{\underset{\displaystyle OH}{\lessgtr}}}}$$

3 5. Write the symbols for the calcium (atomic number = 20) and chloride (atomic number = 17) ions.

4 6. Diagram a hydrogen bond between an $-NH_2$ group and a $-C{\overset{O}{\underset{O^{\ominus}}{\lessgtr}}}$ group.

4 7. Define hydrophobic interactions.

5 8. Define chemical equilibrium for a reversible reaction.

5 9. In the reaction sequence:

$$\text{ADP + Ph} \rightleftharpoons \text{ATP + Cr} \rightleftharpoons \text{CrPh + ADP}$$

What effect would an increase in the concentration of Ph have on the concentration of CrPh? Explain briefly.

6 10. Outline the procedure you would use to prepare 100 ml of a 0.5 molar solution of $KHCO_3$.

Atomic Weights

K = 39
H = 1
C = 12
O = 16

7 11. Explain the meaning of the "pH" of a solution.

7 12. What concentration of NaOH would give a pH of 12.3?

8 13. Write a complete definition or explanation of a buffer system.

9 14. Write the complete structural formula for glucose.

10 15. Match the following:

_____Fructose

_____Ribose

_____Lactose

_____Galactose

_____Maltose

_____One molecule with a beta linkage

_____All disaccharides

_____Any one pentose

A.

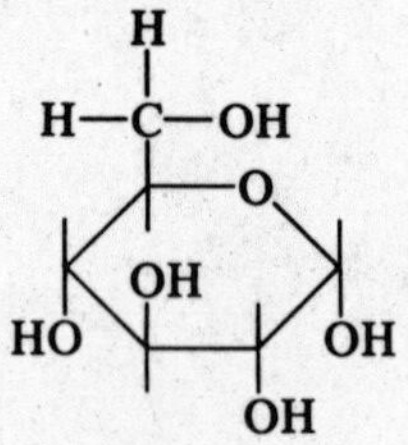

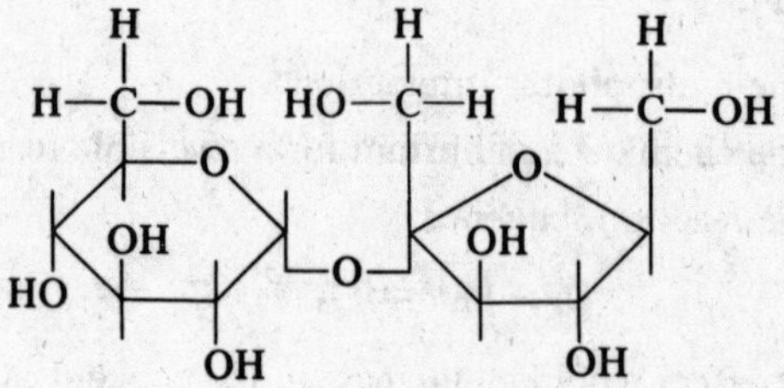

B.

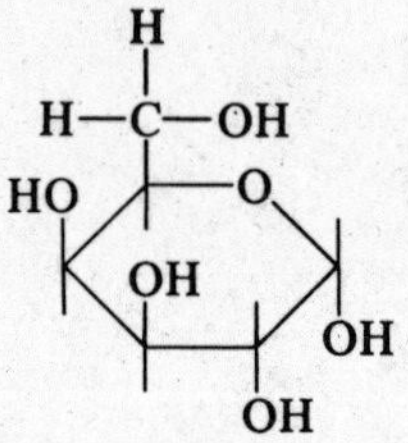

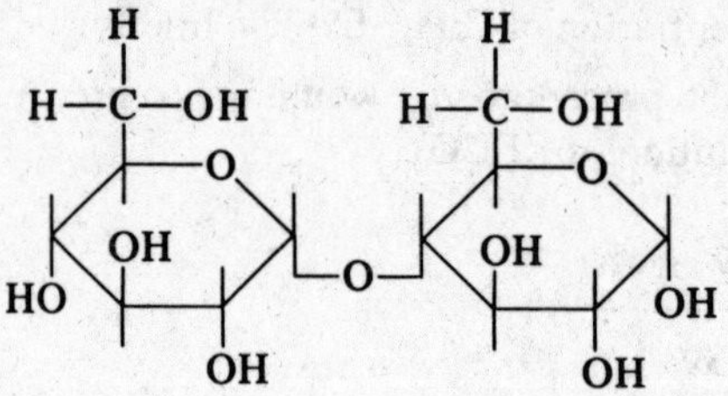

C.

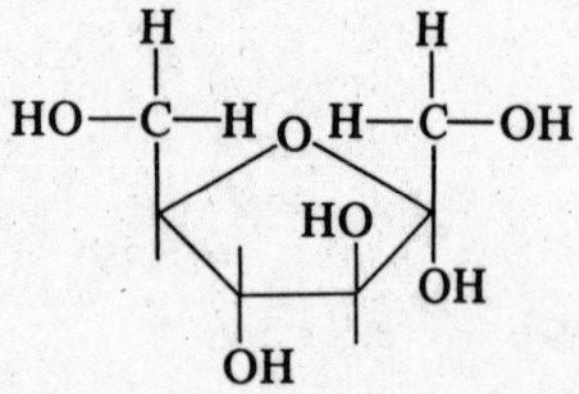

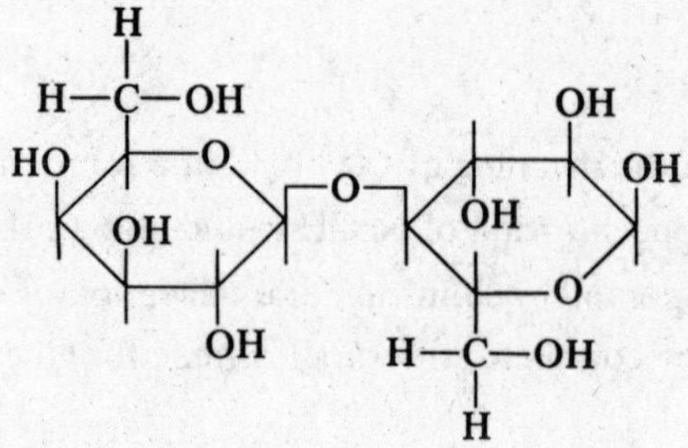

D.

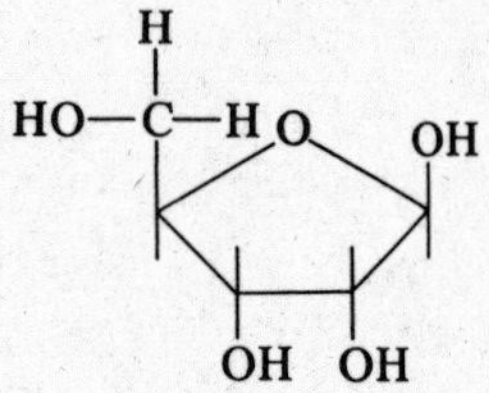

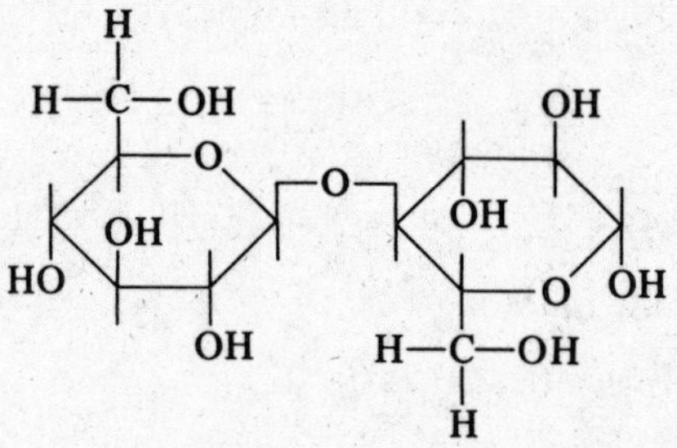

E.

F.

G.

H.

I.

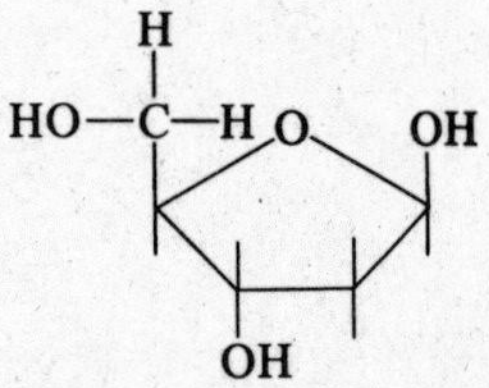

2 16. Write the structural formula for a carboxyl (acid) group.

2 17. Write the structural formula for an ester linkage.

11 18. To what group does each of the following amino acids belong?

___ H_2N–CH₂–CH₂–CH₂–CH₂–CH(NH_2)–COOH

___ H–CH₂–CH(NH_2)–COOH

___ HO–(benzene ring)–CH₂–CH(NH_2)–COOH

___ H–C═C(–N═CH–N–)–CH₂–CH(NH_2)–COOH

___ HO–CH₂–CH(NH_2)–COOH

_____Alanine

_____Cysteine

_____Lysine

_____Phenylalanine

12 19. Diagram the general structural formula of a tripeptide. Circle each peptide unit.

12 20. Describe briefly the structure of the alpha helix.

13 21. Define or explain the primary structure of a protein.

13 22. What is denaturation (briefly)?

14 23. Diagram and label the basic structure of a biological membrane.

14 24. Write the general structural formula of a phospholipid.

14 25. Match the following:

15 _____Purine (Unit 15)

_____Steroid (Unit 14)

_____Pyrimidine (Unit 15)

A.
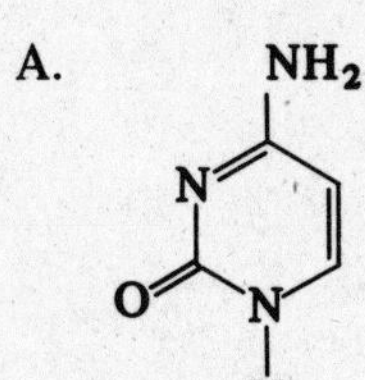

B. OH (phenol ring)

C. (steroid ring skeleton)

D.
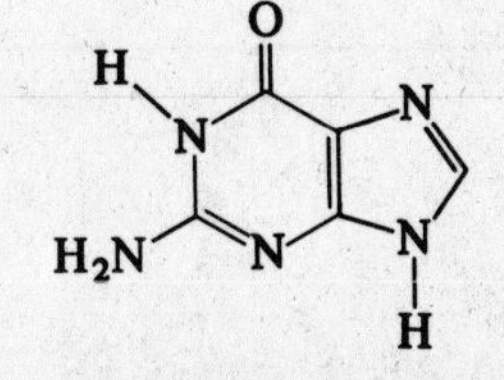

E. (pyrrolidine ring) N–H, –C(=O)OH

15 26. List the three constituents of a nucleotide.

15 27. Construct the RNA strand complementary to this DNA strand (include H bonds).

```
etc. —P—D—P—D—P—D—P—D—P—D— etc.
        |     |     |     |     |
        C     A     T     G     T
```

12 28. _____Who first postulated the structure of the alpha helix?

15 _____Which two persons first postulated the helical structure of DNA?

A. Beadle	E. Pasteur
B. Crick	F. Pauling
C. Darwin	G. Schleiden
D. Linneas	H. Watson

16 29. In the reaction: AMP + Ph + Cal $\rightleftharpoons$ ADP + H_2O

a. At equilibrium will AMP or ADP be present in the greatest concentration? Explain your answer briefly.
b. Is the reaction to the right probably endothermic or exothermic?
c. How many calories does 10 Calories equal?

17 30. Explain the functioning of enzymes briefly.

13 31. What type of protein is usually bonded to chromosomal DNA?

9 32. To what group does each of the following belong? (A test of word endings.)

_____Ribulose (Unit 9)
_____Maltate (Unit 3)
_____Propionic (Unit 3)

CUMULATIVE SELF-TEST KEY

To facilitate scoring, a suggested breakdown of point assignments is given at the right of each correct response. The total points for each response are listed at the left margin. In evaluating your responses, use your best judgment. Remember, your evaluations should be rigorous and should stress the correctness of concepts rather than the use of identical wording. Based on previous posttest results, a score of above 70 is satisfactory and a score above 90 is quite good.

POINTS **RESPONSES**

3 1. ·Mg·
(Mg nucleus: 1 point; electrons: 1 point, position of electrons: 1 point)

3 2.

```
     H  H
     |  |    //O
  H—C—C—C
     |  |    \
     H  OH    OH
```

(1 point off for each error, to a maximum of −3)

5 3. __G__ Aldehyde
__B__ Aromatic
__C__ Ester linkage
__D__ Keto
__A__ Ethyl
(1 point each)

2 4.

```
     H
     |   //O
  H—C—C
     |   \
     H    O⁻ Na⁺
           ↗
```

(Na^{+}—correct position and charge: 1 point; arrow indicating ionic bond: 1 point)

4 5. Ca^{++} Cl^{-}
(1 point each for correct sign, + or −; 1 point for the correct number of charges)

3 6.

```
                 H
                 |
       O···H—N—
     //
   —C
     \
      O⊖
```

(involvement of =O, 1 point; involvement of H, 1 point; use of ··· for the hydrogen bond, 1 point)

3 7. Hydrophobic interactions involve the clustering together of hydrocarbon groups in an aqueous medium.
(1 point each for the mention of hydrocarbons, clustering, and the aqueous medium)

2 8. At equilibrium the rate of the reaction to the right equals the rate of the reaction to the left (the rate of the forward reaction equals the rate of the reverse reaction).
(1 point for the mention of rates and 1 point for mentioning equality)

3 9. The concentration of CrPh would increase. (1 point) An increase in the Ph concentration causes the rate of the reaction to the right to increase; this would cause the concentration of ATP to increase; this would cause the next reaction rate to the right to increase and would increase the concentration of CrPh and ADP. (2 points)
(The explanation might also be based on mass action, equilibrium constants, or a statement of LeChatelier's principle.)

5 10. Molecular weight of $KHCO_3$:

K– 39
H– 1
C– 12
30– 48
100 gram M.W. = 100 grams (1 point)

Calculation: 1 liter of 1 M $KHCO_3$ = 100 g/liter
1 liter of 0.5 M $KHCO_3$ = 50 g/liter
100 ml of 0.5 M $KHCO_3$ = 5 g/100 ml
(2 points)

Procedure: Obtain a 100 ml volumetric flask.
Weigh out 5.0 grams of $KHCO_3$; place in the flask.
Add about 80 ml of distilled water.
Swirl flask until all the $KHCO_3$ is dissolved.
Add distilled water to the 100 ml line, mix, transfer to a lab. bottle and label appropriately. (2 points)

3 11. The pH of a solution is the -log (or log of the reciprocal or log 1/conc.) of the hydrogen ion concentration.
(1 point for mention of the hydrogen ion concentration; 2 points for using -log or log 1/concentration)

5 12. Calculation:

$$pOH = \log \frac{1}{(OH^-)} = \log \frac{1}{(NaOH)} = 1.7$$

pH + pOH = 14
12.3 + pOH = 14
pOH = 14 - 12.3
pOH = 1.7 (2 points)

$$pOH = 1.0 + 0.7$$
$$= \log 10 + \log 5 \text{ (log 5 = 0.7 from memory)}$$
$$= \log (10 \times 5) = \log (0.5 \times 10^2)$$
$$= \log \frac{1}{2} \times \frac{1}{10^{-2}} = \log \frac{1}{2 \times 10^{-2}}$$

$(OH^-) = 0.02$ M (3 points)

(The above methods of calculations are examples. Any method that yields the correct answers is acceptable.)

5 13. A buffer system contains a weak acid plus a salt anion of that acid. The buffer system resists a change in pH (H ion concentration); the H^+ from the acid reacts with excess OH^-, and the salt anion reacts with excess H^+. The buffer system will be most effective when pH = pK (of the weak acid).
(1 point each for mention of the weak acid, salt anion, H^+ from the acid reacting with excess OH^-, salt anion reacting with excess H^+, and pH = pK.

5

14.

```
            H
            |
         H—C—OH
            |
            C—O
      H   /     \   H
       \ /       \ /
        C   OH H  C
      /  \  |  | /  \
    HO    \ C——C     OH
            |  |
            H  OH
```

(1 point each for 6 carbons, carbons in a ring, oxygen in the ring correctly, the position of the OH groups, and the attachment of the CH_2OH group)

8

15.

C	Fructose	G	Maltose
D	Ribose	H or I	One molecule with a beta linkage
H	Lactose	F, G, H, I	All disaccharides
B	Galactose	D or E	Any one pentose

(1 point for each blank: no more than 1 point per item; part wrong = all wrong)

3

16. $-C(=O)O^{\ominus}$ or $-C(=O)OH$

(1 point each for the correct position of –C, =O, and –OH)

3

17. $-C(=O)-O-$

(1 point each for the correct position of –C–, =O, and –O–)

9

18. Basic
Hydrocarbon
OH-polar
N-polar
OH-polar
Hydrocarbon
S-polar
Basic
Hydrocarbon
(1 point each)

3

19.

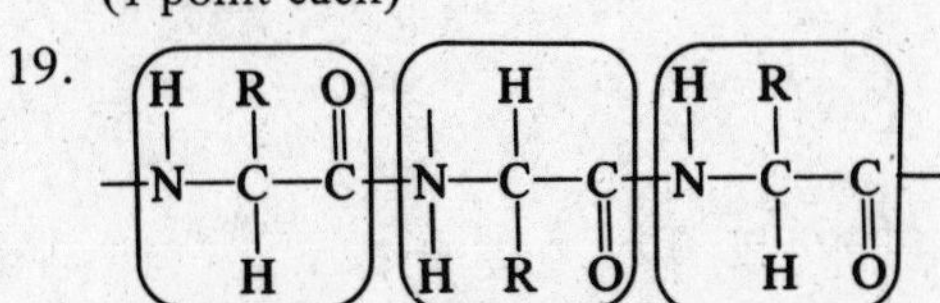

(1 point each for the correct arrangement of the N–C–C etc. chain; =O; and R's and H's. R, H, and =O can be written above or below the N–C–C etc. chain. One point off for each incorrect circle.)

3

20. The alpha helix is a tight coil in a polypeptide (protein) chain. The coil is stabilized by hydrogen bonding between peptide units that are relatively close together (3.6 peptide units) on the chain.
(1 point each for mention of polypeptide or protein, hydrogen bonds, and bonding between close peptide units)

2

21. The primary structure of a protein is the specific sequence of amino acids in the polypeptide chain; it is gene coded (by the sequence of nucleotides in DNA).
(1 point each for mentioning specific sequence of amino acids and gene coding.)

3 22. Denaturation is an alteration of the tertiary structure of a protein resulting in the loss of biological activity. It is frequently due to the disruption of hydrogen bonds. Agents commonly causing denaturation are heat, organic solvents, acids, and bases.
(1 point each for mention of alteration of tertiary structure, disruption of hydrogen bonds, and one or more agents.)

5 23. Phospholipids (1) – About 80% (1)
Lipid bilayer (1)
Sterols (1) – About 20% (1)

(5 points total)

3 24.

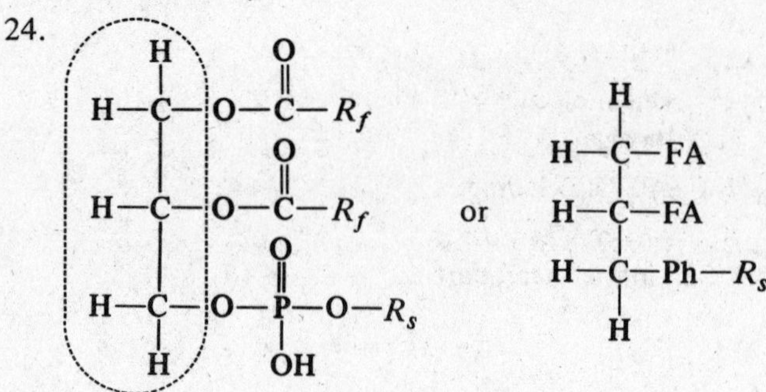

(One point each for glycerol, indicated by the dotted circle; Rf or FA groups; Ph and Rs groups.)

3 25. D Purine
C Steroid
A Pyrimidine (1 point each)

3 26. Phosphate
Pentose sugar
Nitrogen base (purine, pyrimidine) (1 point each)

5 27.

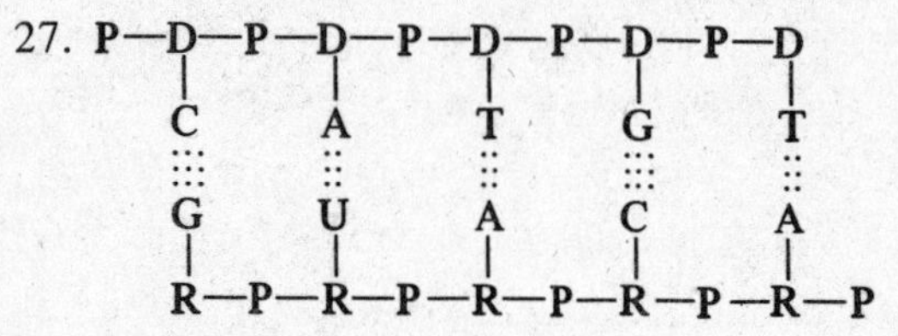

(1 point off for each error up to a maximum total of -5 points.)

3 28. F
B and H (1 point each)

4 29. a. AMP (1 point)
The energy of activation for the reaction to the right is greater than that for the reaction to the left (or similar response.)
(1 point)
b. Endothermic (1 point)
c. 10,000 (1 point)

3	30. Enzymes bind substrate molecules (1 point) in a specific conformation (1 point) that facilitates the formation or breaking of bonds (1 point).
1	31. Basic (histones) (1 point)
3	32. Sugar Salt Acid (1 point each)

(118 total points)

INDEX

Primary reference pages are noted in italics. A citation preceeded by a "T", such as T-14, refers to self-test coverage of the term or concept.